Cambridge Elements ≡

Elements in the Philosophy of Mathematics
edited by
Penelope Rush
University of Tasmania
Stewart Shapiro
The Ohio State University

MATHEMATICS AND EXPLANATION

Christopher Pincock
The Ohio State University

CAMBRIDGE
UNIVERSITY PRESS

Shaftesbury Road, Cambridge CB2 8EA, United Kingdom

One Liberty Plaza, 20th Floor, New York, NY 10006, USA

477 Williamstown Road, Port Melbourne, VIC 3207, Australia

314–321, 3rd Floor, Plot 3, Splendor Forum, Jasola District Centre,
New Delhi – 110025, India

103 Penang Road, #05–06/07, Visioncrest Commercial, Singapore 238467

Cambridge University Press is part of Cambridge University Press & Assessment,
a department of the University of Cambridge.

We share the University's mission to contribute to society through the pursuit of
education, learning and research at the highest international levels of excellence.

www.cambridge.org
Information on this title: www.cambridge.org/9781009017664

DOI: 10.1017/9781009039154

First published 2023

A catalogue record for this publication is available from the British Library.

ISBN 978-1-009-01766-4 Paperback
ISSN 2399-2883 (online)
ISSN 2514-3808 (print)

Mathematics and Explanation

Elements in the Philosophy of Mathematics

DOI: 10.1017/9781009039154
First published online: April 2023

Christopher Pincock
The Ohio State University

Author for correspondence: Christopher Pincock, pincock.1@osu.edu

Abstract: This Element answers four questions. Can any traditional theory of scientific explanation make sense of the place of mathematics in explanation? If traditional monist theories are inadequate, is there some way to develop a more flexible but still monist approach that will clarify how mathematics can help to explain? What sort of pluralism about explanation is best equipped to clarify how mathematics can help to explain in science and in mathematics itself? Finally, how can the mathematical elements of an explanation be integrated into the physical world? Some of the evidence for a novel scientific posit may be traced to the explanatory power that this posit would afford, were it to exist. Can a similar kind of explanatory evidence be provided for the existence of mathematical objects, and if not, why not?

Keywords: explanation, pluralism, platonism, causation, inference to the best explanation

ISBNs: 9781009017664 (PB), 9781009039154 (OC)
ISSNs: 2399-2883 (online), 2514-3808 (print)

Contents

1 Introduction

Some scientific explanations involve mathematics. Within mathematics, some proofs are said to explain. Do these practices tell us anything about the nature of explanation or mathematics? In this Element this daunting topic is divided into four parts. First, can any traditional theory of scientific explanation make sense of the place of mathematics in explanation? Each traditional theory that is discussed is a *monist* theory because it supposes that what makes something a legitimate explanation is always the same (Section 2). Second, if traditional monist theories are inadequate, is there some way to develop a more flexible but still monist approach that will clarify how mathematics can help to explain (Section 3)? After a consideration of the limitations of some recent flexible monist accounts, the options for a pluralist approach are examined. What sort of pluralism about explanation is best equipped to clarify how mathematics can help to explain in science and in mathematics itself? While a pluralist can allow that different sorts of explanations work differently, it still remains important to clarify the value of explanations (Section 4). Finally, how can the mathematical elements of an explanation be integrated into the physical world? Some of the evidence for a novel scientific posit may be traced to the explanatory power that this posit would afford, were it to exist. Can a similar kind of explanatory evidence be provided for the existence of mathematical objects, and if not, why not? (Section 5).

In his 2001 paper "Mathematical Explanation: Problems and Prospects" Paolo Mancosu argues that "mathematical explanations can be used to test theories of scientific explanation and that an account of mathematical explanation might have important consequences for the philosophy of science" (Mancosu 2001, p. 102). This Element builds on this point by considering how various approaches to scientific explanation can make sense of both (i) explanatory proofs in pure mathematics and (ii) scientific explanations that turn essentially on mathematical resources. In Sections 2–4 I argue that the best option for clarifying how these explanations work is pluralism about explanation. This means that different explanations employ different explanatory relevance relations when they indicate why some target is the way that it is. In Section 5 I consider the significance of mathematical explanation for the interpretation of pure mathematics. I argue that the existence of genuine mathematical explanations does not support the existence of mathematical objects through the use of inference to the best explanation.

My own interest in mathematical explanation can be traced directly to the pioneering work of Paolo Mancosu (see especially Mancosu 2000, 2001, 2008, 2018).[1] I was lucky enough to have Mancosu as the advisor for my 2002

---

[1] A new version of (Mancosu 2018) is currently in preparation.

dissertation on questions related to the applicability of mathematics. This Element attempts to follow Mancosu's call to attend carefully to mathematical and scientific practice in philosophical work. I believe that the pluralism about explanation that I argue for is consistent with Mancosu's views, but he may not agree with the account of what all explanations have in common that I offer here (Section 4.3).

A draft of this Element benefitted enormously from comments by Sam Baron, Andre Curtis-Trudel, Marc Lange, and Paolo Mancosu. I have unfortunately not been able to address all of their helpful suggestions here, and they are, of course, not responsible for any remaining errors or oversimplifications of the issues discussed. I am also grateful to two anonymous referees for their insightful reactions to the penultimate version of this Element. I hope this Element will help to introduce new readers to the wonders of mathematical explanation, and also to inspire new work on this complex topic.

2 The Challenge to Traditional Theories of Scientific Explanation

This section starts by introducing five principles that are used to test competing accounts of explanation, and illustrates how these tests work by developing some standard objections to accounts that emphasize derivation and unification (Section 2.1). This section also considers three causal accounts of explanation and argues that they are unable to make sense of the contrast between explanations that merely employ mathematics to represent something else and explanations whose explanatory power is tied more directly to the mathematics employed (Section 2.2).

2.1 Derivation and Unification

Philosophical investigations of a topic like explanation typically take for granted some principles about that topic that make it possible to test competing accounts. This Element takes for granted five principles. It supposes that an account of explanation aims to cover all explanations, including those found in science and mathematics. As I will argue, many accounts fail to respect the principles articulated in this section. We can thus use these principles to identify the problems with various accounts of explanation that have been proposed, especially when one considers how mathematics figures into explanations. Of course, one may avoid these problems by rejecting one or more of the principles that are assumed here.

All participants in these debates agree that a scientific explanation provides a reason why something is the case. The target of an explanation may be something particular, like a specific event or state. The target may also be something general, like a recurring pattern or phenomenon. A legitimate explanation of a target indicates why that target is the way it is. This motivates our first principle for accounts of explanation:

1. There is an important distinction between a description of some target of explanation and an explanation of that target.

This principle does not definitively refute any account, as a defender of any account is liable to interpret "an important distinction" in their own self-serving way. However, I will appeal to this principle to help to clarify my reasons for questioning this or that proposal.[2]

A closely related principle involves the distinction between the evidence that some phenomenon has some character and an explanation of that aspect of the phenomenon. For example, careful paleontological investigations may determine that the rate of the Earth's rotation on its axis is decreasing. But additional accounts of the gravitational interactions between the Earth and the Moon are needed to explain this change. Our second principle is thus:

2. There is an important distinction between the evidence for some fact and an explanation of that fact.

The third principle that I will deploy assumes that there is an order to explanation. If one says that B is the case because of A, then A provides a reason for B being the way that it is. In some respect, then, A must be more basic or fundamental than B. In causal explanation, A is partly responsible for the occurrence of B, and so in this sense is also more basic. If this is right, then it would be illegitimate to appeal to B when explaining A. One way to summarize this point is to say that an explanatory relevance relation is asymmetrical: when A stands in that relation to B, then B does not stand in that very relation to A. However, the issue is complicated by the fact that an explanation may have parts. To allow for explanations with parts, our third principle has the following formulation:

3. (Priority) If A is part of an explanation of B, then B is not part of an explanation of A.

For example, the mass of the Moon is part of the explanation for why the rate of rotation of the Earth on its axis is decreasing over time. Our third principle thus requires that the decreasing rate of rotation of the Earth on its axis is not part of an explanation of the mass of the Moon.

---

[2] For a recent discussion of this issue, see Taylor (forthcoming).

Our fourth and fifth principles help to identify the subject matter of this Element. This is the special character of some explanations that involve mathematics. As we will see later in this section, many philosophers maintain that some explanations that involve mathematics use the mathematics in a special way that renders the explanation genuinely or distinctively mathematical. Following Baker and Baron, I will call such explanations "genuine mathematical explanations."[3] Of course, not everyone agrees that there are genuine mathematical explanations. But our fourth principle takes for granted that genuine mathematical explanations exist and requires that an account clarify their character:

4. There is a special way that mathematics may appear in a scientific explanation that makes it a genuine mathematical explanation.

Our fifth (and final) principle relates to pure mathematics. One goal of mathematical activity is to obtain a proof of a theorem. Mathematicians sometimes praise or criticize a proof based on its explanatory power. In certain contexts, it is thought valuable to explain why a theorem is the case even after it has been given a proof that is otherwise adequate. Our fifth principle assumes that this feature of mathematical practice is legitimate:

5. Some proofs of a theorem explain why that theorem is the case, while other proofs do not explain why that theorem is the case.[4]

Combining our fourth and fifth principles will turn out to be a powerful tool to criticize some proposed accounts of explanation. Many proposals will fail the fourth or fifth test because they do not allow for genuine mathematical explanations or they rule out explanatory proofs. As with the other principles, this does not provide a definitive refutation of these proposals, but it does clarify their limitations and also why some may reject those proposals.

Much of our discussion will turn on cases where mathematics appears in an explanation. Our first case is an answer to the question, "Why is the shadow cast by the Ohio Stadium flagpole 49 m in length at 3 p.m.?" An explanation of this state may appeal to the position of the sun at the time and to the height of the flagpole. But this information does not seem sufficient to explain the length of the shadow, as there is a deductive gap between the statements that provide this information and the statement characterizing the target of the explanation:

---

[3] See especially Baker (2005), Lange (2013), and Baron (2019).

[4] Contrary to the claims of D'Alessandro (2020), nobody assumes that all explanations in pure mathematics are proofs. See especially Mancosu (2001) and Lange (2018b). I restrict my focus here to proofs to make the discussion tractable.

1. At 3 p.m., the light rays from the sun hit the top of the flagpole at an angle of 45°.
2. The height of the flagpole is 49 m.
   Therefore, the length of the shadow is 49 m.

To close this deductive gap, we need to add a statement from geometric optics that involves trigonometry:

3. The length x of the shadow cast by any pole of height y m when the light hits at an angle of 45° satisfies the following equation: $\tan 45° = x$ m$/y$ m.

As $\tan 45° = 1$ and $y = 49$, it follows that $x = 49$ (Figure 1). So for this type of case, at least, the role for the mathematics in the explanation is to permit the deduction of a statement characterizing the explanatory target.

Although the idea has a long history, Hempel is the philosopher who did the most to argue that a necessary condition on an important kind of explanation is that the explanation provide a deduction of a statement describing the explanatory target. Hempel called such explanations "deductive-nomological" (D-N) explanations. The term "nomological" indicates an additional necessary condition on such deductions: they must deduce their target statement through the essential use of a scientific law. Our statement 3 would be the law for this D-N explanation. This is Hempel's way of distinguishing an explanation from a description.

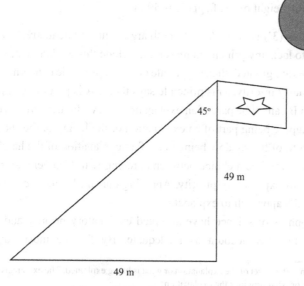

Figure 1 The flagpole and the shadow.

Hempel offered different motivations for the need for laws in explanations. In his famous *Aspects of Scientific Explanation*, for example, Hempel (1965) says "[(i)] the argument shows that, given the particular circumstances and the laws in question, the occurrence of the phenomenon *was to be expected*; and [(ii)] it is in this sense that the explanation enables us to *understand why* the phenomenon occurred." But in addition "[(iii)] it is in virtue of such laws that the particular facts cited in the explanans possess explanatory relevance to the explanandum phenomenon" (p. 337).[5] The relationship between (i), (ii), and (iii) is far from clear. One interpretation of Hempel is that what makes something explanatorily relevant is that this fact could have been used to lawfully predict that state in advance. It is this that constitutes our understanding of that state.

The most influential objection to Hempel's D-N account takes for granted that some laws permit deductions with a troubling sort of symmetry. Our flagpole case was in fact introduced into these debates to illustrate one such troubling case.[6] For in addition to the presumably acceptable explanation just given, the following deductive argument also seems to meet all of Hempel's necessary conditions on D-N explanations:

1. At 3 p.m., the light rays from the sun hit the top of the flagpole at an angle of 45°.
2'. The length of the shadow is 49 m.
3'. The height y of any pole that casts a shadow of length x m when the light hits at an angle of 45° satisfies the equation: $\tan 45° = x \text{ m}/y \text{ m}$.

   Therefore, the height of the flagpole is 49 m.

If (3) is a law, then (3') is also a law. As both arguments are deductively valid, Hempel seems to lack any principled reason to exclude this explanation. But if both explanations are granted, then we violate our third principle concerning the order of explanatory priority. This principle says that if A is part of an explanation for B, then B cannot be part of an explanation for A. But here we have the height of the flagpole being part of an explanation of the length of the shadow, and also the length of the shadow being part of an explanation of the height of the flagpole. Thus we face a choice between agreeing with Hempel and maintaining an order of explanatory priority. Applying our third principle requires rejecting Hempel's approach to explanation.

Most philosophers of science have accepted explanatory priority and thus rejected Hempel's D-N account as inadequate. By far the most popular

---

[5] The explanandum is the target of the explanation or what is being explained. The explanans is the explanation itself or what provides the explanation.

[6] See Salmon (1989) for extensive discussion of this and other objections to Hempel.

approach adds a causal condition on explanation, which we consider in Section 2.2. However, Kitcher offered a different diagnosis of the failings of Hempel's approach. He argued that Hempel failed because he tried to assess explanations individually. The alternative approach that Kitcher pursued is to evaluate explanations globally based on how well they help to unify or system-atize a collection of accepted statements: "Science advances our understanding of nature by showing us how to derive descriptions of many phenomena, using the same derivation again and again, and, in demonstrating this, it teaches us how to reduce the number of types of facts we have to accept as ultimate (or brute)" (Kitcher 1989, p. 432). At any given time in the history of science, there will be some set of accepted statements K. The "explanatory store" over K will specify a set of argument patterns that permit some members of K to be derived from other members of K. An explanation (with respect to this K) will then be an instance of such an argument pattern. Kitcher adds that an explanation is legitimate when it appears "in the explanatory store in the limit of the rational development of scientific practice" (Kitcher 1989, p. 498).

To appreciate the explanatory role for mathematics that Kitcher's approach creates, it will be useful to introduce a case from pure mathematics.[7] As assumed in our fifth principle, only some proofs in mathematics are judged to explain the theorem proven. In Euclid's *Elements* the solution to a problem involves constructing a geometric figure using the limited resources licensed by his postulates, for example to connect any two points by a line or to draw a circle around a point with the radius of some line. One such problem is to bisect an angle BAC (Book I, proposition 9; see Figure 2).

The first step to solve this problem is to pick some point on line BA. Call this point D. A circle centered on A and of radius AD can then be drawn to cross line CA at a new point E. Radius AD is equal in length to AE. An earlier construction in Euclid shows how to construct an equilateral triangle on any given line. Construct such a triangle on line DE, with a third corner F. Finally, connect F to A. The triangles ADF and AEF are congruent, as they have three sides of the same length (AD = AE, DF = EF, AF = AF). As congruent triangles have corresponding angles of the same size, angle DAF = angle EAF, and the bisection is completed. Let us suppose that this construction not only proves that every angle can be bisected, but also that it explains why every angle can be bisected.

A construction procedure like the bisection of an angle can be iterated, and so it is clearly possible to divide an angle into n equal parts when n satisfies the equation $n = 2^m$, for some m ($n = 2, 4, 8, 16, \ldots$). Discussing a case like this,

---

[7] For Kitcher's own examples from pure mathematics, see Kitcher (1989, p. 424).

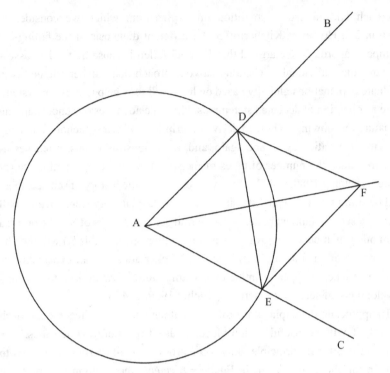

Figure 2 Bisecting an angle.

Kitcher notes that "[e]ven when we are interested in explaining a particular event or state, the explanation we desire may well be one that would also explain something quite general, and any attention to the local details may be misguided and explanatorily inadequate" (Kitcher 1989, p. 427). In this case, one may ask why an angle can be divided into 64 equal parts. While one proposed explanation would lay out all the steps of the construction, Kitcher's view is that a better explanation would connect the 64-part case to all the cases that are amenable to a unified construction procedure. The goal of unification is often prominent in pure mathematics as well as in many scientific cases. In Kitcher's terms, this would lead the explanatory store over a K that includes Euclidean geometry to contain a single argument pattern that covers all divisions of angles into n parts, where n is a power of 2. The instances of this pattern would then count as explaining their respective theorems.

More generally, the explanatory store over K is arrived at by identifying the smallest number of stringent argument patterns that permit the most members of K to be conclusions of some instance of some such pattern.[8] It is not clear if

---

[8] A stringent argument pattern places substantial conditions on how its instances can be generated. These conditions are needed to avoid making unifications too easy to achieve.

Kitcher's procedure for picking out the explanatory store over K is well defined, and there are difficult issues associated with how this procedure is meant to accommodate scientific change. Kitcher does offer an intriguing place for mathematics in explanation, though. For it does seem that mathematical theorems are well suited to unify large numbers of mathematical and nonmathematical claims. This is clear even from our two cases so far. For both the flagpole case and the bisection case, the generalizability of the mathematical result shows how many similar cases can be treated in a uniform fashion. All the members of K that involve a shadow being of a certain length can be handled using a single argument pattern whose crucial premise generalizes (3): The length x meters of the shadow cast by any object of height y meters when the light hits at an angle of z degrees satisfies the equation tan z degrees = x meters / y meters. Similarly, for any number of parts $n = 2^m$, the complete instructions for how to bisect any angle into that many parts can be given using $(n - 1)$ iterations of the angle bisection construction.[9]

Kitcher also proposed an ingenious way to preserve explanatory priority using his unificationist approach to explanation (Kitcher 1989, p. 484). His general strategy was to argue that any purported explanatory store E(K) over K that allowed for troubling symmetrical pairs of derivations would have redundant argument patterns. So, E(K) could be replaced by a different explanatory store E'(K) that would provide a better unification of K by disqualifying one of the proposed explanations. In the flagpole case, Kitcher allows for an argument pattern that derives the height of flagpoles from lengths of shadows using our generalization of (3). But he insists that there will be another argument pattern that derives the dimensions of ordinary objects like flagpoles and towers in terms of their origin and development or, as we might put it, their constitution. The height of the flagpole can be derived by summarizing how it was constructed, so that its parts combine to yield an object of this height. If E(K) has this argument pattern and also has an argument pattern H that derives the height of the flagpole based on the length of the shadow it casts, then argument pattern H only derives statements that can also be derived in some other way. Thus, an explanatory store E'(K) that drops pattern H and retains the constitution pattern would mark an improvement.

The powerful role of mathematics in unifying derivations turns out to be a problem for Kitcher's approach. We can see this by recalling our first two principles for an account of explanation: a description is not an explanation, and evidence for a target is not an explanation of that target. Kitcher is preoccupied

---

[9] As each bisection increases the number of parts by 1, dividing an angle into $n = 2^m$ parts requires $n - 1$ bisections.

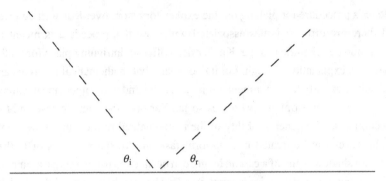

Figure 3 Law of reflection.

with the number of stringent argument patterns needed to derive a given set of claims. Here is a case from geometric optics that Kitcher's analysis gets wrong.[10] The law of reflection says that when a ray of light hits a reflecting surface like a mirror, the angle of incidence θ_i equals the angle of reflection θ_r (Figure 3).

Another law of optics is Snell's law, which concerns refraction: when a light ray goes from one medium (like air) to another (like water), the direction of the ray will change or refract. If the new medium is more dense than the old medium, the angle that the ray makes to the normal axis will decrease so that $\theta_1 > \theta_2$ (Figure 4).

Snell's law connects the ratio of the sines of these angles to the ratio between the so-called refractive indexes n_1, n_2 of these media:

$$n_1/n_2 = \sin \theta_2/\sin \theta_1.$$

The measured refractive indexes of different media were found to increase with the density of the media, but no further explanation for why this law obtained was apparent.[11]

If we follow Kitcher, then we should adopt an argument pattern as explanatory if it permits one to treat reflection and refraction together.[12] One such argument pattern deploys Fermat's principle that a light ray will travel between two points on the path that minimizes the time of the trip. Both the law of refraction and Snell's law can be derived in a uniform fashion from Fermat's principle. The derivations take the endpoints of the light ray's path to be fixed and vary the point O at which

---

[10] More involved examples of the same kind are used to criticize Kitcher in Morrison (2000). See also Hafner and Mancosu (2008) for criticisms of Kitcher's proposal for explanatory proofs.

[11] See Nahin (2004) for more on this case.

[12] This assumes a set of statements K where there is no way to obtain an explanatory store over K that lacks this pattern and that scores better on Kitcher's criteria. Arguably, such a K was present during Fermat's time, prior to our current understanding of the nature of light.

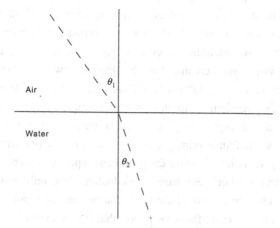

Figure 4 Law of refraction.

the ray strikes the surface. Basic trigonometry then suffices to calculate the time for each potential path. For any reflection, the least-time path is the symmetric path required by the law of reflection. For Snell's law, one must assume that the speed of light in some medium, v_i, is determined by the ratio between the speed of light in a vacuum, c, and the refractive index of that medium, n_i ($v_i = c / n_i$). Qualitatively, the travel time of the light will be less if it travels longer in the medium with a higher speed of light. More mathematically, the time for each potential path can be surveyed by varying the point O where the ray crosses media. The least-time path can then be identified and found to fit Snell's law.

The objection to Kitcher's unification approach to mathematical explanation is that this uniform derivation does not explain the instances of these laws. One diagnosis of this failure is that a uniform derivation of the instances of these laws may not clarify the processes involved. There is no reason to suppose that the light ray somehow chooses a path that minimizes the travel time. Minimizing the travel time is not a reason why light travels in this way. As we might put it, we have arrived at a uniform, mathematical redescription of the phenomena of reflection and refraction, but this is not yet an explanation for why light follows these laws. It seems that Kitcher's approach to explanation ignores this distinction between uniform redescription and explanation.[13]

One lesson, then, is that many mathematical derivations are not explanations, but redescriptions. Our principle of the order of explanatory priority also highlights that many mathematical derivations are not explanations, but merely predictions. There is a third kind of case to consider, though. We might have

---

[13] As Reutlinger (2018) notes, Hempel claimed that Fermat's principle could be used to explain. I consider Reutlinger's reasons for endorsing this aspect of Hempel in Section 3.1.

cases where the mathematics is present in a legitimate explanation but the mathematics itself is not explanatory. One way to parse such cases is that the mathematics is there functioning merely to represent something else that is doing the explaining. But there might still be other cases where the mathematics is doing the explaining. In "Mathematical Explanation by Law" Baron calls this the "genuineness problem": "to uphold a meaningful distinction between explanatory and non-explanatory uses of mathematics in science" (Baron 2019).[14] This is our fourth principle for an account of explanation. Baron's plan (in this paper) for how to solve the genuineness problem involves building on Hempel's and Kitcher's emphasis on deduction. But, unlike Hempel and Kitcher, Baron considers the different ways that mathematics can afford derivations. In one kind of case, the information that the mathematics contributes concerning the explanatory target is exhausted by how the mathematics is representing the target. Consider, for example, a case where a mathematical claim conveys information about a physical system by first picking out an abstract mathematical structure and then relating it to the target by some kind of structural mapping. Our flagpole case is like this, where different geometric lines are interpreted in terms of light rays, physical heights of objects, and lengths of shadows. For such cases, it looks like we have only a mathematical representation of some nonmathematical explanation. Baron summarizes this kind of descriptive information this way: "M contains descriptive physical information relevant to [target] P when M contains information about an aspect of some mathematical structure that is *mapped into* the physical structure corresponding to P" (Baron 2019, p. 709, emphasis added). But Baron argues that in other cases the mathematical claims provide information about the target over and above what is represented via these mappings: in such cases M provides information about P "in virtue of carrying information about an aspect of a mathematical structure and yet *that aspect is not a relatum in a mapping relation between the mathematical and physical structures at issue*" (Baron 2019, p. 710).

My worry about Baron's approach is that information can be provided in this way, and so not be descriptive in Baron's sense, and yet still fail to be explanatory. To help readers appreciate this worry, I will first summarize a case where it looks as if Baron's test works. Consider the seven bridges that at one time connected four areas of the city of Königsberg (Figure 5).

It turns out that it is impossible to make a tour of these bridges that crosses each bridge exactly once. One explanation of this impossibility relates the bridges to an abstract mathematical graph (Figure 6).

---

[14] See Baker (2005).

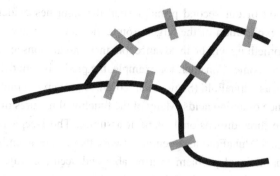

Figure 5 The bridges of Königsberg.

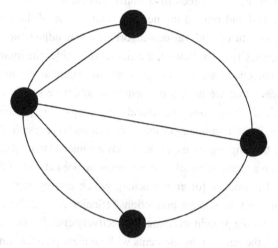

Figure 6 A graph.

This relation is one of the structural mappings that Baron talks about: bridges are mapped to edges, and areas of land are mapped to nodes. This mapping associates paths across the bridges with sequences of connected edges. We can show that this abstract graph has no sequence that proceeds from node to node along edges such that each edge is included exactly once. This entails that it is impossible to cross all the bridges exactly once on a tour of the bridges. But this impossibility is "not a relatum in a mapping relation between the mathematical and physical structures at issue." So Baron's test is met, and we have an explanatory mathematical derivation of a feature of a physical system. In these cases, "the structural mappings constitute a point of contact across which non-descriptive information may be conveyed" (Baron 2019, p. 709).

The problem with Baron's proposal is that we can specify cases where Baron's test is met and yet these derivations are not explanations. That is, we have a violation of our second principle that distinguishes evidence for the target and an explanation of the target. Sometimes our evidence allows us to predict how something will be in advance, but these predictions are not explanations of that outcome. Consider, for example, the predictive model for protein folding known as AlphaFold (Senior et al. 2020). A protein is a molecule made up of a sequence of amino acids. Many of the functional features of the protein are tied to the three-dimensional shape it assumes. The DeepMind research group developed AlphaFold as a neural network that can efficiently solve this daunting computational problem in a reliable and accurate way. The neural network contains a large number of elements whose connections transform the input information about some given sequence of amino acids into a representation of the three-dimensional structure of the protein. The researchers trained and refined the network using some of the available data about protein structures and various algorithms that adjust the connections between elements. Crucially, though, the elements (besides the input and output layers) and connections are not assigned any determinate physical interpretation. This means that we have a mathematical structure that is almost completely detached from any structural mappings to protein features. Unsurprisingly, in their summary of this very successful research, the scientists do not employ the language of explanation, saying instead that "[h]ere we show that we can train a neural network to make accurate predictions of the distances between pairs of residues [of amino acids], which convey more information about the structure than contact predictions" (Senior et al. 2020, p. 706). The information about the protein structure is effectively encoded in the mathematical structure of the network by elements with no clear physical interpretation. So Baron's test is met, but this mathematical derivation does not explain why that protein has that structure. Baron might respond that a derivation of protein structure using AlphaFold is not an explanation because the derivation lacks some other necessary features of an explanation. I would agree with Baron on this point. But this shows that we cannot identify what makes something a genuine mathematical explanation by focusing on how mathematics affords derivations.[15]

2.2 Causation

Already we can see that some approaches to explanation face severe difficulties. I will say that an approach is a monist approach if it supposes that every explanation involves the very same explanatory relation. I think of these relations as relations between facts in the world: if A obtains because of B, then

---

[15] See Povich (2021, pp. 516–519) for a similar worry, along with concerns of a different kind.

A stands in some special explanatory relevance relation to B.[16] Our genuineness principle assumes that some cases have mathematics in the explanation, and yet the mathematics is not standing in the appropriate explanatory relation because it is representing the explainers. A monist approach makes it hard to see what the mathematics is doing in such explanations. More importantly, many relations that monists examine seem to preclude mathematics from standing in that relation to targets. So, a monist risks ruling out genuine mathematical explanations altogether (Zelcer 2013; Kuorikoski 2021).

This challenge is easy to appreciate for the main versions of a causal account of explanation. A causal account first says what causes are and then clarifies how explanations relate to causes. In broad strokes, the three main approaches to causal explanation can be traced back to Salmon, Lewis, and Woodward. Salmon argued that causes are a type of process (Salmon 1984). So we can explain something by describing a causal process that led up to its occurrence. Salmon had a somewhat restricted notion of a causal process tied to physics, but more recently the so-called new mechanists have extended the basic approach so that it is more flexible. One summary statement of this mechanist approach is "Mechanisms are entities and activities organized such that they are productive of regular changes from start or set-up to finish or termination conditions" (Machamer et al. 2000, p. 3). Our flagpole case would then involve a mechanism associated with the transmission of light. Mathematics can function to represent mechanisms, but it is only the mechanisms themselves that are explanatory. The attempted explanation of the height of the flagpole via the length of the shadow then fails because there is no mechanism that is "productive" in this direction.

David Lewis' influential work on causal explanation aspires to characterize causes in terms of counterfactuals. One simple proposal is that event C causes event E just in case that were C not to have occurred, then E would not have occurred. However, this simple proposal is vulnerable to a number of counterexamples. Suppose S throws a rock (C) that breaks a window (E), but that were S not to have thrown her rock, then B would have thrown his rock (C') and broken the window (E). In such a case, C causes E, and yet it is not the case that were C not to have occurred, then E would not occur, as were C not to have occurred, the backup C' would have occurred and brought about E. To handle this worry and other objections Lewis required that a cause be connected to an effect through a series of intermediate events in a special way: small counterfactual variations in the cause go along with small counterfactual variations in some intermediate events and in the effect (Lewis 2004, p. 91). For example, S's throwing of the rock, the actual cause, is connected to the breaking by intermediate events such as

---

[16] See Pincock (2018) for more discussion of this approach.

the rock being halfway to the window. In addition, small changes in S's throwing of the rock go along with small changes in how the window breaks. By contrast, small changes in B's throwing make little to no difference to how the window breaks. Lewis' hope was that we can single out the right events as causes using these elaborate collections of counterfactuals.

Lewis' counterfactual approach has no difficulties with our flagpole case, as small changes in the height of the flagpole do go along with small changes in the length of the shadow. How does Lewis rule out the opposite direction as noncausal and so nonexplanatory? Consider the claim that were the shadow to have been shorter, then the flagpole would have been shorter. To evaluate this counterfactual we consider the possible situation that is most similar to the actual situation, but where the shadow is shorter. One initial puzzle is whether we should consider a possible situation where the shadow has been made shorter through a change in the flagpole. If that situation is appropriately similar to the actual world, then the counterfactual claim would come out true. This kind of counterfactual is known as a "backtracking" counterfactual: it becomes true by considering a situation where the truth of the consequent is the basis for the truth of the antecedent.[17] Other backtracking counterfactuals are tied to predictions ("If the barometer reading had not fallen, then the storm would not have occurred") or identities ("If water was not a liquid at room temperature, then H_2O would not be a liquid at room temperature"). A counterfactual approach to causal explanation needs to rule out some counterfactuals as explanatory if it is to satisfy our first three principles for description, evidence, and priority.

Lewis argued that backtracking counterfactuals are not the right kind of counterfactual for an analysis of causation. He eliminated backtracking counterfactuals from his analysis of causation by comparing possible situations to actual situations in a special way. The counterfactuals needed for causes have their antecedents made true in a way that involves a special sort of miraculous, minimal departure from the actual world. This minimal change preserves as much as possible concerning the intrinsic features of the actual situation along with the actual physical laws. Given how the world is, one such miraculous, minimal departure would shorten the length of the shadow by inserting a new light source that left everything else as it actually is. Such a minimal change in the length of the shadow would not go along with a change in the height of the flagpole. That is, the counterfactual "if the shadow were shorter, then the flagpole would have been shorter," as Lewis intended it, comes out false. So, the length of the shadow is not a cause of the height of the flagpole.

---

[17] For a counterfactual of the form "If A were the case, then B would be the case," A is the antecedent and B is the consequent.

Woodward also emphasizes the links between causation and counterfactuals, but winds up with a very different account than Lewis (Woodward 2003). One source of disagreement concerns the issue about how to rule out backtracking counterfactuals. Woodward rejects Lewis' appeal to miraculous, minimal departures from the actual world as too metaphysical. First, it is hard to know how to implement this notion of a minimal departure in actual scientific cases, given the central role of the intrinsic character of objects. Second, scientists do not seem to worry about this issue, which suggests that there is a scientific notion of causation that can be developed without Lewis' apparatus. For Woodward, the best strategy is to make causation relative to a set of variables. A variable is a family of values, where we suppose that an object has at most one of these values at a time. In the flagpole case, we consider a variable set that includes the height of the flagpole H, the angle that the sun makes θ, and the length of the shadow L. Other variables in the set might include the ambient temperature, wind speed, and color of the flagpole. To see if H is a cause of L, we need to consider what Woodward calls an intervention on H. This is a special sort of causal change in H from its actual value that holds the value of other potential causal variables in the set fixed. If an intervention on H brings about a change in L, then we will say that H is a cause of L. There is no intervention on L that brings about a change in H, for we are only permitted to make an intervention by changing L using some outside change, and not by changing H. So, unlike Lewis' appeals to miraculous changes and similarities between situations, Woodward argues that he can get the right account of causes by tying being a cause to a choice of variable set. The appeal to interventions makes Woodward's account nonreductive: we appeal to a special sort of causation when we talk about interventions, and so there is no way to specify whether or not C is a cause of E without using causal notions. Woodward argues that the account is still fruitful, as it clarifies the nature of causes in a way that makes clear how evidence for the existence of causes can be assembled.

One puzzle for these three causal approaches to explanation is to clarify the importance of the mathematics to the explanation when the mathematics is merely representing causes.[18] All three views of causal explanation can point to the powerful way that mathematics affords a description of what they take causes to be. For a mechanist, the mathematics can track how a mechanism develops over space and time, and so provide a compact and surveyable

---

[18] Perhaps a more significant problem for causal approaches is our fifth principle: some mathematical proofs explain. If we assume that causes are absent from pure mathematics, then a causal approach is in trouble. Pluralist approaches to explanation that allow for explanatory proofs are considered in Section 4.

representation of what is making some target behave in the way it does. For Lewis and Woodward, the same point holds for the counterfactuals that they place at the center of an explanation. Woodward often emphasizes the powerful way that a mathematical relationship between the values of his variables can be exploited to present a causal generalization that covers many systems. This allows a scientist to answer many "what if things had been different" questions using the right kind of counterfactual. These generalizations differ in their scope of application and range of invariance. Other things being equal, Woodward claims that generalizations with a wider scope or range of invariance contribute to better explanations. So the mathematics is not just used to represent causes, but may be a means to arrive at a better explanation than would be otherwise available by representing these causes in some other way.

In all these cases what explains some target is causes. No causes are mathematical in themselves or intrinsically, but for various cognitive purposes we may find it convenient to represent those causes using the mathematics we have available. Some mechanists put this point by emphasizing the ontic or worldly character of explanation. As Craver argues, "Conceived ontically … the term ['explanation['] refers to an objective portion of the causal structure of the world, to the set of factors that produce, underlie, or are otherwise responsible for a phenomenon. Ontic explanations are not texts; they are full-bodied things" (Craver 2014, p. 40). Not all mechanists are comfortable singling out this ontic notion as basic, and the counterfactual approaches of Lewis and Woodward also do not need to go as far as Craver does. But the basic orientation of all these causal approaches is to restrict the explanatory role of mathematics to the representation of causes, the legitimate explainers.

This brief survey should make it clear why a causal approach to explanation struggles to make sense of genuine, nonrepresentational cases of mathematical explanation. For a causal view, if the mathematics is not itself causing the target phenomenon, then it is not, strictly speaking, explaining. It is thus very difficult for a causal approach to satisfy our fourth principle that genuine mathematical explanations exist. If the defender of a causal approach to explanation wishes to maintain their monism about explanation, then they must deny the existence of genuine mathematical explanations. I take this to be a significant cost of causal monist views as there certainly seem to be genuine mathematical explanations like the bridges case discussed in Section 2.1.[19]

---

[19] See Woodward (2018) for a discussion of how Woodward's approach can be generalized to include noncausal explanations. Lange (2021a) raises some concerns for this approach.

2.3 Conclusion

In this section I have considered six strict monist accounts of explanation that include mathematics. Hempel, Kitcher, and Baron had difficulties with our first three principles tied to description, evidence, or explanatory priority. The causal monist views of the new mechanists, Lewis and Woodward made all uses of mathematics in explanation representational. So, if we insist on our fourth principle that some mathematical explanations are not representational, then we must adopt a different approach to explanation.

3 Counterfactual Accounts of Mathematical Explanation

A monist about explanation who wants to include both causal explanations and genuine mathematical explanations must somehow generalize their explanatory relevance relation so that it covers both sorts of cases. One approach involves treating a mathematical claim like a causal law: we consider variations to the actual world and use the mathematical claim, like we would use a causal law, to indicate what else would change. This is the core idea of Reutlinger's generalization of a causal counterfactual approach to explanation (Section 3.1). Another approach treats a mathematical fact as more like a cause: we should consider variations in the mathematical fact itself and what results from this variation. This is just like how, in the causal case, we consider what would happen were a potential cause to be different. I consider three versions of this approach in Section 3.2. The problems with both approaches to generalizing a counterfactual approach motivate the discussion of pluralist views in Section 4.[20]

3.1 Counterfactuals Involving Only Possible Worlds: Reutlinger

In a number of papers Reutlinger has developed a monist "counterfactual theory of explanation."[21] As he puts it, "A monist holds that causal and non-causal explanations share a feature that makes them explanatory" (Reutlinger 2018, p. 77). This leads to a theory that takes an especially simple form: something counts as an explanation just in case it meets four conditions that Reutlinger calls Structure, Veridicality, Inference, and Dependency. Structure requires that the explanation include a "nomic generalization" such as a law and some additional statements relating to initial or boundary conditions. Veridicality mandates that these statements be true or at least approximately true. To satisfy Inference, the statements in question must entail the statement to be

---

[20] See Lange (2021a), Kostić and Khalifa (2021), Lange (2022a), and Kasirzadeh (forthcoming) for other discussions of the limitations of counterfactual approaches.

[21] Reutlinger (2016), Reutlinger (2017), Reutlinger (2018), and Reutlinger et al. (2022).

explained.[22] Finally, Dependency requires that the nomic generalization support a counterfactual with a special form: "if the initial conditions . . . had been different than they actually are (in at least one specific way deemed possible in the light of the nomic generalizations), then E [the target of the explanation] . . . would have been different as well" (Reutlinger 2018, p. 79). A nomic generalization supports a counterfactual when the procedure for evaluating the counterfactual leads to the counterfactual being true, on the assumption that the nomic generalization is true. Reutlinger helpfully illustrates how this works for a number of examples, including the bridges case we discussed in Section 2.1. The nomic generalization in this case is the mathematical theorem that a graph G has the right sort of path just in case "(i) all the nodes in G are connected to an even number of edges, or (ii) exactly two nodes in G (one of which we take as our starting point) are connected to an odd number of edges" (Reutlinger 2018, p. 84). One counterfactual that is supported by this theorem is "if all parts of Königsberg had been connected to an even number of bridges, then people would not have failed to cross all of the bridges exactly once" (Reutlinger 2018, p. 84). The truth of this counterfactual means that Dependency is satisfied. As the other three conditions that Reutlinger imposes are also satisfied, the bridges case counts as an explanation.

Reutlinger's counterfactual tests for scientific explanations do not involve an alteration of any mathematical fact. We need only consider ordinary changes in the initial or boundary conditions. In the bridges case, this requires considering only some physically possible alternative configuration of bridges. One worry with requiring only these sorts of counterfactuals is that it risks failing our fourth principle, namely respecting the distinction between genuine mathematical explanations and explanations where the mathematics is representing nonmathematical explainers. Reutlinger's monism effaces this distinction, as his counterfactual test works in the same way for an ordinary case like the flagpole case and in a mathematical case like the bridges case. So, if we require our fourth principle, then we can conclude that Reutlinger's monism is untenable.

To be fair, it should be clear that Reutlinger does not claim that he can recover a special sort of genuine mathematical explanation. Instead, Reutlinger focuses on the contrast between causal and noncausal explanation. Here Reutlinger breaks from Woodward and argues that causal explanations should not be identified in terms of the interventions that Woodward requires. Reutlinger develops a different approach to causal explanation that he traces back to

---

[22] Reutlinger also allows an entailment of a conditional probability of the target E given the initial/boundary conditions, provided that this conditional probability is greater than the prior probability of the target. I will set this aspect of Reutlinger's account aside, as it is not relevant to the objections I will raise.

Russell: if an explanation exhibits four features, then it counts as causal, while if any of these four features are absent, then the explanation is noncausal. The four features are asymmetry, time asymmetry, distinctness, and metaphysical contingency (Reutlinger 2018, p. 89). The flagpole case has all four of these characteristics, so it counts as a causal explanation. By contrast, the bridges case violates both distinctness and metaphysical contingency. To see why, suppose someone claimed that the mathematical theorem and mapping caused the failure to make the right kind of circuit on the bridges. Reutlinger objects that we have too close a relationship here for one fact to cause the other: "the explanandum fact that people fail to cross each bridge exactly once supervenes on (or metaphysically depends on, or is grounded in) the explanans fact that Königsberg instantiates a particular kind of graph (and the fact that people actually attempted to cross the bridges)" (Reutlinger 2018, p. 91). However, causes and effects do not stand in this sort of supervenience relation. So the mathematics with the mapping is not a cause of the bridge crossing failure. Similarly, causes are thought to be metaphysically contingent. For example, in the flagpole case, the causal process could have worked quite differently. But for the bridges, Reutlinger sees a metaphysical necessity: "It is metaphysically, or mathematically, impossible (and not merely physically impossible) to cross the bridges as planned, if Königsberg instantiates an non-Eulerian graph" (Reutlinger 2018, p. 91). Again the conclusion is that we have a noncausal explanation.

This negative strategy for identifying noncausal explanations has the same limitations that we saw earlier for Baron's 2019 proposal: noncausal explanations are hard to distinguish from predictions (Baron 2019, section 1.1). Consider again the predictive model for protein folding, AlphaFold, that we introduced in Section 2.1. As we saw, this model is highly accurate and suffices to derive a wide range of protein structures. This means that Reutlinger's first three conditions of Structure, Veridicality, and Inference are satisfied. The only question, then, is whether or not the model provides nomic generalizations that support the kind of counterfactual that Reutlinger specifies in his Dependency condition. Consider a particular protein: hemoglobin. The AlphaFold model is able to predict the shape of hemoglobin in both the normal case and the case of the genetic variant that produces sickle cell anemia. At the level of amino acids, these two proteins differ only in one place, where a "glutamic acid residue" in the normal protein is replaced by a "valine residue" in the variant responsible for sickle cell anemia (Ingram 1957, p. 326). As the model can predict the shape of both proteins on the basis of their different amino acid chains, it does support a counterfactual of the kind that Reutlinger requires. For any instance of a sickle cell anemia hemoglobin, the model entails that were this protein to have a glutamic acid residue in place of the salient valine residue, then it would develop a different shape, namely the shape of normal hemoglobin. Reutlinger's additional conditions on causal explanation do not seem to be met, but for Reutlinger this only means that we have a case of noncausal explanation. I conclude that Reutlinger has lost the crucial difference between noncausal explanation and evidence. That is, he has failed to satisfy our second principle.

Reutlinger may embrace this result, as he also endorses a number of controversial claims about explanation, at least for noncausal explanation. For example, Reutlinger allows violations of explanatory priority: if A is part of an explanation for B, then B cannot be part of an explanation for A (Section 2.1). Whenever there is a nomic generalization that is a biconditional of the form "A if and only if B," it will support pairs of counterfactuals of the form "Were not-A, then not-B" and "Were not-B, then not-A."[23] So, in such cases, as long as Reutlinger's other three conditions are met, we will have pairs of explanations that violate explanatory priority. Reutlinger here singles out the bridges case for this sort of symmetry. We have not only the counterfactual noted earlier, but also this one: "If people were able to cross all of the bridges exactly once, then all parts of Königsberg would be connected to an even number of bridges, or exactly two parts of town would be connected to an odd number of bridges" (Reutlinger 2018, p. 92). Reutlinger is not troubled by this result. I conclude that, as with Hempel, if we wish to preserve explanatory priority, then we should not adopt Reutlinger's counterfactual approach.

Reutlinger also embraces Hempel's attitude toward the case of Fermat's least time principle explanations (Reutlinger 2018, pp. 82–83). As usual, Structure, Veridicality, and Inference are clearly satisfied, as we may derive both the law of reflection and Snell's law of refraction using the least time principle, and the principle (if formulated carefully) is true. Reutlinger notes how various counterfactuals arise from Fermat's principle that involve what would be the case if the endpoint of the light ray had been varied. For example, "if the beam had traveled from point A at t_1 to point B* at t_3 (in contrast to point B), the beam would have gone through point C* at t_2 (in contrast to point C)" (Reutlinger 2018, p. 83). So, if Reutlinger is right, then Fermat's derivation does count as an explanation. It will just be a noncausal explanation. Recall that our reason for bringing up this case was to show how Kitcher failed our first principle that distinguished between description and explanation. Reutlinger has the same problem. We can either follow Reutlinger and count a wide range of

---

[23] Reutlinger is clear that the main feature of a nomic generalization is that it supports counterfactuals. See Reutlinger (2018, p. 79, fn. 7).

mathematical descriptions as explanations or else abandon this counterfactual theory in order to preserve the difference between mathematical description and explanation.

3.2 Counterfactuals Involving Impossible Worlds

A second strategy for generalizing a counterfactual analysis of causal explanation is to suppose that mathematical facts stand in the same relation to their targets that a cause stands to its effect. With this analogy in mind, we should consider counterfactuals where that very mathematical fact is altered just as we consider counterfactuals where the actual cause is altered. This requires counterfactuals that alter the truth of some mathematical claims. As most assume that these truths are necessary truths, there will be no possible world as usually conceived to consider. The three proposals we discuss in this section avoid this by invoking "impossible worlds" so that these mathematical counterfactuals have nontrivial truth-values.

3.2.1 Baron, Colyvan, and Ripley's Counterfactual Approach

A generalization of Lewis' counterfactual account of causal explanation has been pursued along these lines by Baron, Colyvan, and Ripley (BCR) in two papers, "How Mathematics Can Make a Difference" (Baron et al. 2017) and "A Counterfactual Approach to Explanation in Mathematics" (Baron et al. 2020). The same counterfactual approach is pursued in both articles, but the 2017 article focuses on mathematical explanations of physical phenomena, while the 2020 article discusses mathematical explanations that are internal to mathematics. I will develop their proposal using two examples that we have discussed so far: the bridges case and the angle bisection case (Section 2.1). For the bridges, one key mathematical claim is (M1) that if a graph has four nodes and each node has three or five edges, then there will not be a sequence of edges that connects nodes and that includes each edge exactly once. In the angle bisection case, a central mathematical claim is (M2) that an equilateral triangle can be constructed on any line segment. Baron, Colyvan, and Ripley's proposal is that these mathematical claims are part of the explanation of their respective targets just in case a counterfactual with a special character comes out true. The counterfactual says that were this mathematical claim *not* to be true, then the target phenomenon would *not* occur. That is, were M1 to fail for graphs, then there would be a circuit of the bridges, and were M2 to fail for line segments, then there would not be a bisection construction.

These counterfactuals for mathematics raise a number of puzzles concerning their interpretation and how to evaluate them. Baron, Colyvan, and Ripley generally follow Lewis in providing a recipe for how to isolate the situation involved: there is an account in terms of similarity between the actual and nonactual scenarios that pins down what to consider. In a case like the bridges case, they summarize the procedure as follows:

> First, hold fixed the morphism [mapping] between the mathematical structure S that appears in the counterfactual and the physical structure P. Second, make a change in the mathematics while holding fixed as much as one can without inducing a contradiction. Finally, consider the ramifications of the change by looking at the way(s) in which the physical structure P twists in response to the twiddling in S in order to preserve the morphism. (Baron et al. 2017, p. 10)

The "twiddling" here is a change in the mathematical structure S, for example one function is substituted for another function. In the pure mathematics cases, the mapping between the mathematical structure S and the physical structure P is no longer relevant. For these cases BCR emphasize the aim of preserving the intrinsic features of the mathematical entities: "In the mathematical case, as in the non-mathematical case, this [evaluation procedure] means holding fixed as much as we can concerning the intrinsic properties of whatever mathematical features are mentioned in the antecedent of a given counterfactual, compatible with realising the antecedent itself" (Baron et al. 2020, p. 7). For example, a line segment's straightness should be held fixed, although how it intersects other lines may change.

In the bridges case, then, suppose that M1 is false so that, for some graph with these nodes and this number of edges, there *is* a sequence that connects vertices and that includes each edge exactly once. We cannot visualize this sort of mathematical impossibility, but BCR claim we are still able to consider what would be the case in such an "impossible world." If the mapping between the graph and the actual bridges is preserved through this change in the character of the graph, then it seems that the bridges must also be "twisted" or altered in character by this change in the mathematics. Given that in the world we are considering there is a sequence that connects nodes and that includes each edge exactly once, and the mapping that takes edges to bridges is maintained, in this situation there *will* be a circuit of the bridges that crosses each bridge exactly once. So the specified counterfactual comes out true, and the mathematical claim is genuinely explanatory.

In the bisection case, suppose that M2 is false, so that it is not in general possible to construct an equilateral triangle on any line segment. This means that the crucial step in the construction is not generally available. If we could establish that in such an impossible mathematical scenario there was no alternative available means of bisecting the angle, then that would show that in this scenario, some angles could not be bisected. For BCR this would then establish that the claim M2 was genuinely explaining the possibility of bisecting any angle. One way to specify this scenario would be make the plane "gappy" so that for some line segments there was no point in the plane that would allow the equilateral triangle to be constructed.[24] In such a plane our construction of the bisected angle would fail. If no other construction worked, then some angles would be unbisectable. And arguably we have not changed the intrinsic character of the objects mentioned in the mathematical claim, such as line segments and angles.

3.2.2 Baron's Counterfactual Schemes

The BCR account should satisfy the monist about explanation because the very same kind of counterfactual test is applied to determine if either some purported cause or mathematics is standing in this single explanatory relevance relation. However, it does not seem that the BCR account is viable. One powerful objection to the BCR proposal has been developed by Baron himself in a 2020 paper, "Counterfactual Scheming" (Baron 2020). We can relate this objection back to Baron's derivational proposal from 2019 (Section 2.1).[25] Baron's basic worry is that a simple counterfactual test will not discriminate genuine mathematical explanations from cases where the mathematics is merely representing some explanatory subject matter such as a cause. We can appreciate this worry using our original flagpole case. A central mathematical claim here is (M3) when a right-angled triangle has an angle z, the length of the side opposite the angle divided by the length of the side adjacent to the angle is tan z. According to BCR, this mathematics will explain just in case that were M3 to be false, then the length of the shadow would be different. Applying the procedure outlined by BCR leads to the conclusion that this counterfactual is true. For were M3 to be false, tan z would have to be different than the actual ratio. Enforcing the mapping originally in place entails that the length of the shadow would be different. In general, keeping the mapping in place transfers the upshot of the change in the mathematical structure over to the physical world. But Baron correctly points out that this transfer is too indiscriminate. It ignores the important contrast between genuinely explaining and merely representing.

In this 2020 paper Baron does not rely on the derivational test from the 2019 paper that we criticized in Section 2.1. Instead, he proposes a combination of a counterfactual approach with a unification approach inspired by Kitcher.

---

[24] For example, a line segment centered at the origin of the xy-plane with endpoints at $(-1,0)$ and $(1,0)$ requires a point at either $(0, \sqrt{3})$ or $(0, -\sqrt{3})$ for the construction of an equilateral triangle. If a plane lacked these points, then no equilateral triangle could be constructed.

[25] Baron also makes this objection to counterfactual accounts of mathematical explanation in an earlier paper (Baron 2016, p. 11).

On this combined approach, for a mathematical claim to figure in an explanation, we need to consider counterfactuals involving mathematics like the ones identified by BCR. However, only some of these counterfactuals render the original mathematical claim genuinely explanatory. Two additional features must be present. The first feature is unification: the counterfactual must be an instance of a "counterfactual scheme" CS1, all of whose instances are true and where some instances involve physical systems that are governed by different physical laws. Baron calls such systems "nomically distinct." The second feature is used to distinguish genuine from merely representational uses of mathematics: this counterfactual scheme CS1 is not superseded by a nonmathematical counterfactual scheme CS2 that covers all the same cases at CS1. For CS2 to disqualify CS1's instances from explaining, CS2 must cover all these cases by mentioning the very physical features that are changed through rendering the antecedent of the counterfactual scheme CS1 true. Using these more demanding requirements, Baron hopes to isolate the genuinely explanatory uses of mathematics from the merely representational uses of mathematics.

To see what Baron has in mind, let us return to the flagpole case. It has Baron's first feature, as there are many systems that exhibit the triangular pattern exhibited by the flagpole and its shadow, and many of these systems are nomically distinct. For example, the length of the base of an inclined plane stands in the very same ratio to the height of the inclined plane that we have in the case of the length of the shadow and the height of the flagpole. However, the second feature is not present: there is a nonmathematical counterfactual scheme that covers all these cases. Consider a parallel counterfactual scheme that mentions ratios between lengths instead of numbers. The way that this nonmathematical counterfactual scheme covers its cases is by bringing about changes in the ratios between the lengths of the objects making up these nomically distinct physical systems. The explanatory power of the mathematical counterfactual scheme is thus undercut by this physical counterfactual scheme. This gives us the right result for the flagpole case: it is not a case of genuine mathematical explanation, but only a case where mathematics represents physical explainers.

Baron calls his latest proposal the "u-counterfactual theory" (Baron 2020, p. 25). One point to note about this theory is that it distinguishes between genuine and representational uses of mathematics by abandoning strict monism about explanation. The genuine mathematical explanations obtain in virtue of an elaborate relation spelled out in the u-counterfactual theory. But ordinary causal explanations of physical phenomena involve causes standing in a different relation to their targets. If Baron's account is right, then we must abandon strict monism in favor of a more flexible kind of monism that requires only a more generic relation between explanations. Baron himself frequently expresses the hope for such a generic theory, noting that "a counterfactual approach to extra-mathematical explanation opens up the enticing possibility of developing a theory of scientific explanation that is fully general" (Baron 2020, p. 26).

One worry about Baron's u-counterfactual theory is that it seems to rule out all cases of genuine mathematical explanation and so fails our fourth principle. To make the problem clear, let us suppose what is sometimes called a supervenience principle: there is some basic physical level for any system such that a change in any feature of the system will go along with a change in that basic physical level of the system. If this principle holds, then all features of the system supervene on this basic level in the sense that there is no change in the features of the system without a change on this basic level. If we accept this supervenience principle, then Baron's second feature for genuine mathematical explanation will never be present: for any mathematical counterfactual scheme, there will be a nonmathematical counterfactual scheme that undercuts the explanatory power of the mathematical one. For all we need to do is consider what is going on at the basic physical level that makes our supervenience principle obtain. This will provide a nonmathematical counterfactual scheme involving just this level that matches the mathematical counterfactual scheme. So there will be no genuine mathematical explanations of physical phenomena.

Baron might respond that a nonmathematical counterfactual scheme need not merely cover these systems, but also unify them in some way that matches (or improves on) the unification afforded by the mathematical counterfactual scheme. This response involves taking the explanatory virtue of unity to be decisive not just in the quality of an explanation, but in its very existence. Defending such a strong claim about unity would require arguments that I do not think Baron has yet provided. We will return to this issue in Section 5. Our puzzle will be how advocates of mathematical explanation should deal with this issue. In Section 5.3 I argue that unity is not the right kind of feature to focus on when considering the legitimacy of an explanation.[26]

I claim that this supervenience issue raises a serious dilemma for Baron's u-counterfactual account. If he accepts supervenience, then there will be no genuine mathematical explanations. But if he rejects supervenience, he faces a new worry that is similar to our objection to Kitcher. Recall that Kitcher struggled to satisfy our first principle, namely to respect the difference between a unified mathematical redescription of a family of phenomena and the

---

[26] See also Knowles (2021b).

explanation of that family of phenomena (Section 2.1). Our example involved Fermat's least time principle. It can be employed to arrive at a unified derivation of both the law of reflection and the law of refraction. It can also be used to derive patterns that involve how light behaves due to other physical laws such as the laws offered by Einstein's general theory of relativity.[27] The least time principle involves a mathematical formula taking on the least value, given the constraints on the system, but the character of these constraints varies across cases. Baron must admit, then, that there is an available counterfactual scheme that involves the possibility that this mathematical formula has some different character. Applying Baron's procedure to this case makes the instances of this schema true: were this formula to have a different value, then light would not conform to the laws in question. So, it looks like Baron counts this sort of mathematical redescription as a legitimate explanation, just as with Kitcher. If Baron endorsed a supervenience principle, he could, of course, rule out this case, but only at the cost of ruling out all genuine mathematical explanations. Alternatively, if Baron gave up supervenience, he could no longer respond that there was guaranteed to be some level of physical description at which a nonmathematical counterfactual scheme that covered all these cases could be specified. And so he would not be able to rule out this derivation as a case of genuine explanation. That is, Baron's u-counterfactual account fails either our first principle or our fourth principle.[28]

3.2.3 Povich's Ontic Counterfactual Account

The last monist, counterfactual account that I will discuss has recently been offered by Povich (2021).[29] Povich motivates his proposal by setting out three features that an account of genuine mathematical explanations should have: we need to respect the (i) modal import, (ii) distinctiveness, and (iii) directionality of these explanations. By modal import Povich just means that genuine mathematical explanations explain some physical phenomenon by showing it to be necessary. We saw this in the bridges case, which Povich also emphasizes: it is impossible for anyone to make this circuit of the bridges in virtue of the relation that the bridges stand to the mathematical graph. Distinctiveness is Povich's

---

[27] The least time principle can be used to derive how light will travel near massive bodies and to describe the resulting phenomena of "gravitational lensing." As one discussion puts the point, "This angle [of deflection] is usually derived by finding the shape of a null geodesic in the Schwarzschild geometry surrounding a spherically symmetric star . . . Here, we have derived the same result using Fermat's principle instead, a complementary approach" (Nandor and Helliwell 1996, p. 48).

[28] Baron (forthcoming) provides yet another proposal that he calls the "Pythagorean" account. This account is briefly discussed in Section 5.2.

[29] See also Povich (2020; forthcoming).

term for what is special about genuine mathematical explanations as opposed to explanations where mathematics represents some other sort of explainer.[30]

Povich's directionality requirement is similar in spirit to explanatory priority, but somewhat more demanding. Recall that priority requires that if A is part of an explanation of B, then B is not part of an explanation of A. Directionality involves pairs of putative explanations with a more intricate relationship. Consider the particular network structure of the actual bridges of Königsberg, where one node has five edges and the other three nodes have three edges (Section 2.1, Figures 5 and 6). Call this network structure P. The explanation that we want to endorse is that the bridges' possession of network structure P explains why there is no circuit of a particular kind that, following Povich, we will call an Eulerian walk. However, Povich does not want to endorse a "reversed" companion to this explanation that involves a different situation. Consider a different town L where their bridges have a network structure Q that has two vertices with three edges and two vertices with two edges (Figure 7).[31] Why do the bridges in town L fail to have network structure P? A proposed explanation of L's failure to have bridges with network structure P is that these bridges permit an Eulerian walk. Povich's directionality requirement says that if the first explanation targeting Königsberg's bridges is legitimate, then this second explanation targeting L's bridges is illegitimate.

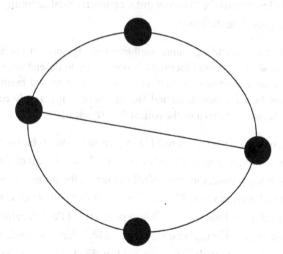

Figure 7 The bridges of town L.

---

[30] Povich uses Lange's term "distinctively mathematical explanation." Povich also frames his proposal as a response to Lange's modal account of these explanations. I will postpone a discussion of Lange until Section 4, as Lange is a pluralist about explanation.

[31] Povich describes the companion as also involving Königsberg, but I have brought in a different town for ease of exposition.

I will regiment Povich's directionality requirement as a principle that restricts how statements that ascribe related features to different objects can appear in legitimate explanations. Let "A(x)" and "B(x)" be two sentence types with a variable x whose tokens have the variable replaced by names. Then directionality says that, for any c, d, if A(c) is part of an explanation of B(c), then it is not the case that not-B(d) is part of an explanation of not-A(d). In the legitimate explanation c = Königsberg, A(c) is "The bridges of Königsberg have network structure P," and B(c) is "The bridges of Königsberg do not permit an Euler walk." In the illegitimate explanation, d = L, not-B(d) is "The bridges of L do permit an Euler walk," and not-A(d) is "The bridges of L do not have network structure P."[32]

For this pair of cases, Povich's directionality principle is very plausible. While the Königsberg explanation clarifies why the Euler walks are impossible, the proposed L explanation does nothing to clarify why the bridges of L fail to have network structure P. As Povich puts it, these failed explanations "give quite good reason to believe their conclusion, but they do not explain it" (Povich 2021, p. 516). This is the contrast involved in our second principle: evidence for some target does not explain that target. I agree with Povich that some violations of directionality lead to violations of our second principle. My main objection to Povich's proposal is that his emphasis on counterfactuals also leads to violations of directionality.

Povich calls his proposal a "narrow ontic counterfactual account" or NOCA. The official version is as follows:

> an explanation is a DME [genuine mathematical explanation] just in case either (a) it shows a natural fact weakly necessarily to depend counterfactually only on a mathematical fact, or (b) it shows a natural event to be necessitated by a component natural fact that weakly necessarily counterfactually depends only on a mathematical fact. (Povich 2021, p. 526)

I will restrict my discussion to clause (a), which turns out to be the more basic case for Povich. Povich's natural facts are states of affairs involving concrete objects. In the bridges case, the state of affairs is that the bridges of Königsberg do not permit an Eulerian walk. The mathematical fact is that network structure P does not permit an Eulerian walk. This explains what Povich calls the narrow target of explanation: "Königsberg's bridges, which have network structure P, do not permit an Eulerian walk" (Povich 2021, p. 527). This is the narrow target as opposed to the wide target "Königsberg's bridges do not permit an Eulerian walk." The narrow target includes a specification of the structure of the bridges, and this information is used to pick out the right sort of counterfactual scenario

---

[32] See also Kostić and Khalifa (2021) for a similar analysis of directionality and its implications for what they call topological explanations.

when checking for counterfactual dependence. As with BCR and Baron (2020), Povich mandates that we consider a special counterfactual involving the mathematical claim: "Had network structure P permitted an Eulerian walk, Königsberg's bridges, which have network structure P, would have permitted an Eulerian walk" (Povich 2021, p. 527). But unlike BCR and Baron (2020), Povich imposes a more demanding requirement than just the truth of the counterfactual. To have an explanation, Povich requires that this counterfactual holds in every possible world where these bridges exist. That is, the counterfactual is not just true of the actual world, but weakly necessarily true.[33] When (and only when) the counterfactual linking the mathematical fact to the state of affairs (narrowly construed) is weakly necessarily true, we have a genuine mathematical explanation of this state of affairs.

Povich claims that his approach is better than other counterfactual accounts, as his insistence that the counterfactual be weakly necessarily true allows him to respect directionality. For example, consider the proposed explanation for why the bridges of L do not have network structure P. It invokes the very same mathematical fact that network structure P does not permit an Euler walk and the additional fact that the bridges of L do permit an Euler walk. If we shift to the narrow target of explanation, we get that the bridges of L, which permit an Euler walk, do not have network structure P. The mathematical claim that network structure P does not permit an Euler walk does entail this target, so we can see right away that Povich's directionality requirement rules out entailment as a sufficient condition on explanation.[34] However, the salient counterfactual is not weakly necessarily true: "Had network structure P permitted an Eulerian walk, [L]'s bridges, which permit an Eulerian walk, would have had network structure P" (Povich 2021, p. 527). This counterfactual fails to be true in any world where the mathematical structure is changed, but L's bridges have some structure besides P that permits an Eulerian walk. As we have seen, L's actual bridge structure is not P, but Q. There is no reason to think that changing the mathematical structure would alter L's actual bridge structure. So, plausibly, this counterfactual is not even true of the actual world. More generally, in cases that threaten to violate directionality, Povich argues that the counterfactual will fail to be weakly necessarily true.

The difference between the legitimate case and the illegitimate case is easy to appreciate once we recall that many network structures permit Euler walks, and many other network structures do not permit Euler walks. So while appealing to

---

[33] A claim about any contingent object like the bridges is not strongly necessarily true, as the claim will not come out true in worlds where the object fails to exist.

[34] Povich uses this argument to object to Baron's derivational proposal (Baron 2019) that we considered back in Section 2.1.

network structure P is apt to explain a failure to complete an Euler walk, appealing to the completion of an Euler walk is not apt to explain the absence of network structure P. Povich's somewhat intricate proposal is to be credited for capturing this important contrast in counterfactual terms. However, as Povich realizes, we can specify cases that satisfy all of Povich's conditions and yet violate directionality. I claim that his response to this problem is not adequate and calls into question his monist counterfactual proposal.

We can obtain violations of directionality for a counterfactual proposal whenever the mathematical claim used to explain is a biconditional. These violations involve negating the biconditional and considering the upshot for the appropriately narrowed target. Such a biconditional is available in the case of networks that permit an Euler walk: as noted earlier, a network structure permits an Euler walk just in case either zero or exactly two nodes have an odd number of edges. This pattern of edges is realized in various different ways in different graphs, but I will suppose it is a legitimate mathematical property. I will call this property "being near even." So our mathematical claim is that a network structure permits an Euler walk just in case the network is near even. This entails that a network structure fails to permit an Euler walk just in case the network is not near even. Using these mathematical biconditionals, we can get violations of directionality using the bridges of Königsberg and city L. Let $A(x)$ and $B(x)$ be the same as before except that we replace "having network structure P" with the more abstract property "not being near even": c = Königsberg, $A(c)$ is "The bridges of Königsberg have a network structure that is not near even," and $B(c)$ is "The bridges of Königsberg do not permit an Euler walk." In the other case, d = L, not-$B(d)$ is "The bridges of L do permit an Euler walk," and not-$A(d)$ is "The bridges of L have a network structure that is near even." For the Königsberg case, we first alter $B(c)$ so it is the narrow target that Povich emphasizes: "The bridges of Königsberg, which have a network structure that is not near even, do not permit an Euler walk." Now consider a counterfactual that involves our mathematical biconditional and this target: "Had network structures that are not near even been just those that permitted an Eulerian walk, Königsberg's bridges, which have a network structure that is not near even, would have permitted an Eulerian walk." This counterfactual is weakly necessarily true, and so according to Povich's account we have a genuine mathematical explanation of the narrowed target. In addition, we have the narrowed target for the bridges of city L: "The bridges of L, which do permit an Euler walk, have a network structure that is near even." Now the associated counterfactual to consider is "Had the network structures that do permit Euler walks been just those that are not near even, the bridges of L, which do permit an Euler walk, would have had a network structure that is not near even." This counterfactual is also weakly necessarily true, and so according to Povich's account we have another genuine mathematical explanation of a narrowed target. But together the two cases violate directionality, as we have a token of A(x) being part of an explanation of a token of B(x) and also a token of not-B(x) being part of an explanation of a token of not-A(x).

While Povich does not consider this exact worry, he does realize that he must place some restrictions on the mathematical claims that can figure in explanatory counterfactuals. As he puts the point, "Excluding some reversals [violations of directionality] . . . requires NOCA firmly to commit to the idea that some mathematical objects exist, and others do not (and cannot). This means that NOCA is committed to a privileged mathematical ontology – not just anything goes" (Povich 2021, p. 537). It does seem right to insist that genuine mathematical explanations will rely on some privileged mathematical ontology.[35] However, Povich's account mandates ruling out mathematical properties like being near even simply because they are part of mathematical theorems that have a biconditional form. By contrast, mathematical practice often takes these properties to be significant, in part because they do explain. To take a simple example, Euclidean triangles are just those plane figures whose interior angles sum to 180 degrees. We cannot exclude either of these properties from our mathematical ontology simply because, if we adopt Povich's account, it leads to violations of directionality. Given the ontology that we find in mathematical practice, if we accept directionality, then we must reject Povich's account.

One could preserve a monist counterfactual account and standard mathematical ontology by giving up directionality. This is consistent with explanatory priority, but would involve awkward combinations of explanations that differed only in the objects involved, as with Königsberg and city L discussed previously.[36]

3.3 Conclusion

This chapter has considered four different flexible monist proposals that generalized counterfactual analyses of causal explanation to include some noncausal cases. The counterfactual theory developed by Reutlinger made do with possible worlds, but as a result allows for too many noncausal explanations that violate our principles concerning description, evidence, and priority. Baron, Colyvan, and Ripley (Baron et al. 2017, 2020; Baron 2020) and Povich (2021) went further and invoked impossible worlds where some mathematical truths

---

[35] We will return to this point in Section 4.

[36] A more thorough investigation of the connections between priority and directionality is not possible here. See again Kostić and Khalifa (2021) for an analysis of this issue in terms of their non-ontic approach to topological explanation.

come out false. Here I objected that the accounts either failed to meet our four principles or, in the case of Povich, failed Povich's own directionality constraint. The upshot of the limitations of these proposals is the rejection of monism, even in its more flexible form. While some explanations will turn on the truth of some sort of counterfactual, other explanations will involve a different explanatory relevance relation. I consider the prospects for this kind of pluralism in the next section.

4 Explanatory Pluralism

A pluralist about explanation supposes that the best way to make sense of the variety of explanations in science and mathematics is to posit two or more explanatory relevance relations. As noted in Section 2, I take these relations to connect facts in the world when one fact explains another. The views we will consider in this section accept a notion of causal explanatory relevance. But each adds a new kind of explanatory relevance that will aid in the clarification of genuinely mathematical explanations. However, we will begin our discussion of pluralist views by considering explanations from pure mathematics. In these cases some mathematical facts explain another mathematical fact. The pluralist expansion of relevance relations avoids the problems that monist views ran into, but it also raises a new problem. Is there anything that unites explanations if they involve these very different explanatory relations? In Section 4.3 I will consider some features that all explanations might have in common, even if pluralism about explanatory relevance obtains.

4.1 Explanatory Proofs in Mathematics

4.1.1 Steiner on Explanatory Proofs

If we examine mathematical practice, then we find that mathematicians value some proofs because those proofs not only show that a theorem is true, but also explain why the theorem is true. This is our fifth principle from Section 2.1. Steiner offered a pioneering analysis of these explanatory proofs.[37] As we will see in Section 4.2.1, Steiner also used this analysis to clarify what we are calling genuine mathematical explanations in science. The basic assumption of Steiner's approach to explanatory proof is that "to explain the behavior of an entity, one deduces the behavior from the essence or nature of the entity"

---

[37] Another analysis of explanatory proof was offered by Kitcher and considered in Section 2.1 (Kitcher 1989). See Hafner and Mancosu (2008) for a critical examination of Kitcher's analysis. See also Mancosu (2001), Pincock (2015b), D'Alessandro (2021), and Ryan (2021) for additional proposals and cases that are too involved to develop here. One attempt to analyze explanatory proofs in terms of counterfactuals was briefly considered in Section 3.2.1.

(Steiner 1978a, p. 143). However, Steiner claims that these essences are not available for mathematical objects. The next best thing is a "characterizing property" of that object: "a property unique to a given entity or structure within a *family* or domain of such entities or structures" (Steiner 1978a, p. 143), where the notion of family is taken as basic. We can illustrate characterizing properties by returning to Euclid's proof that any angle can be bisected (Section 2.1). As we saw, this proof can be extended into a family of proofs that any angle can be divided into n equal parts if, for some m, $n = 2^m = 2, 4, 8, 16, \ldots$. If we are considering a family of angles divided into equal parts, then a characterizing property of the objects in this family is how many parts the division has. Then, one of our proofs shows how a bisected angle is available using a particular construction, while an angle divided into four equal parts is available using a closely related construction.

Steiner's proposal is that an explanatory proof invokes a characterizing property in a way that allows one to vary that characterizing property to arrive at other proofs concerning other objects in the domain: "an explanatory proof depends on a characterizing property of something mentioned in the theorem: if we 'deform' the proof, substituting the characterizing property of a related entity, we get a related theorem" (Steiner 1978a, p. 147). Deforming a proof may involve a kind of mechanical manipulation, or it could be something more creative that preserves the original "proof-idea" (Steiner 1978a, p. 147). Our collection of angle division proofs satisfy Steiner's criteria. For we can see how shifting from a bisected angle to an angle with four equal parts requires an extended construction. So we have a family of proofs, where appropriate shifts in the characterizing properties go along with new, successful constructions.

One objection to Steiner's proposal is that there are nonexplanatory proofs that meet his tests. The case I will focus on is the classification of finite simple groups. A group is a type of object from abstract algebra that has found application throughout mathematics and physics. The finite simple groups are (in a sense) the building blocks of all the other finite groups, just as the prime numbers are the building blocks of all the other natural numbers. After a great deal of effort, an exhaustive classification of the finite simple groups was achieved in the 1980s. The resulting proof of the "Classification Theorem" showed how every finite simple group was one of 29 identifiable types.[38] Three of these types were expected, but the remaining 26 types are called "sporadic" as they fail to exhibit any clear connections to one another or to the three expected types. This has led some practitioners to conclude that even though there is a classification of these groups, we do not yet have an explanation for

---

[38] See Solomon (2001, p. 341) for a compact statement of the theorem.

why the finite simple groups come in just these kinds. As one mathematician puts it: "Are the finite simple groups, like the prime numbers, jewels strung on an as-yet invisible thread? And will this thread lead us out of the current labyrinthine proof to a radically new proof of the Classification Theorem?" (Solomon 2001, p. 346).[39] I take this to mean that the current proof fails to explain the Classification Theorem. The thought seems to be that the classification is haphazard and unilluminating. Lange describes such cases as involving what he calls a mathematical coincidence (Lange 2017).[40] A classification will appear to involve a coincidence when "a salient feature of the result is its identifying some property common to cases that otherwise seem unrelated" (Lange 2017, p. 280). As we will discuss later in Section 4.1.2, part of Lange's view is that an explanatory proof shows that an apparent coincidence is really not a coincidence. Our worry, then, is that Steiner must reject this plausible link between explanations and noncoincidences: classifications that strike practitioners as involving coincidences can still meet all of Steiner's tests for mathematical explanation.

The root of the problem is that the classification allows us to prove, for any group that is appropriately characterized, whether or not it is a finite simple group. A characterizing property of a group would be the structure of the group, namely the number of objects in the group and the network established by the group operation. As this structure is varied, we can easily shift to proofs that establish whether or not the group in question is a finite simple group. For all we have to do is check whether or not the group falls into one of the 29 types of finite simple groups. So, all of Steiner's conditions are met. I conclude that Steiner's conditions make it too easy to arrive at an explanatory proof.

Another objection to Steiner's proposal is that there are explanatory proofs that fail his tests. One sort of problematic case is noted by Steiner at the end of his article (Steiner 1978a, pp. 148–150).[41] Suppose we investigate domain X by relating the entities in X to the entities in some other domain Y. Mathematicians often do this by abstracting away from some of the differences in X so that many objects in X are assigned the same object in Y. Examining the objects in Y can still explain what is going on in X. But, as Steiner concedes, Y does not afford characterizing properties of the objects in X, so his official test for explanatory proof cannot be met. All he says here about the worry is that "the concept of 'characterization' will have to be weakened to allow for partial characterization"

---

[39] See also Thompson's remarks from 1982, given at Solomon (2001, p. 346). Additional history and discussion may be found in Ronan (2007).

[40] See also Baker (2009a).

[41] See Hafner and Mancosu (2005) for another kind of case that Steiner's account misses.

(Steiner 1978a, p. 150). However, if Steiner amends his tests in this way, he will have to count even more nonexplanatory proofs as explanatory.

To appreciate the problem, I will sketch a case involving Euclidean constructions that is widely viewed as explanatory. It turns out that while any angle can be bisected using Euclid's postulates, there are some angles that cannot be *tri*sected using these postulates. One such angle is the 60-degree angle found on each corner of an equilateral triangle. More generally, for some n, any angle can be divided into n parts, while for the remaining n, some angles cannot be divided into n parts. An explanatory proof of this result should indicate what it is about these numbers that is responsible for the associated construction being always available or sometimes impossible.

Here is how the explanatory proof works (Stewart 2004, chapters 6, 7). First, associate a Euclidean construction with the constructability of points in the Euclidean plane. The basic idea is that a point is constructable just in case its coordinates may arise as solutions to a series of a restricted class of polynomial equations. For example, a point with the coordinates $(0, \sqrt{2})$ is constructable because the square root of 2 is a solution to the equation $x^2 = 2$. If we start with the field of rational numbers, adding the square root of 2 produces what is called a field extension. Each field extension can be assigned a degree using the degree of the polynomial equation required to introduce the new elements to the field. So, in our example the degree of the field extension is 2. The key result of the proof is then that a point is constructable just in case it falls within a field extension whose degree is some power of 2 (Stewart 2004, p. 79). Finally, we can characterize n such that any angle can be divided into n equal parts. For $n = 2$, the field extension has the required feature, but for $n = 3$, the field extension fails to have the right feature. This is why some angles cannot be trisected.[42]

The problem for Steiner is that this explanatory proof does not provide a characterizing property of a given construction. For example, trisecting an angle can be treated using a field extension, but many other constructions may also be treated in the same way. So, whenever we embed one domain X into another domain Y using this sort of many–one mapping, we will not get the characterizing properties of the objects in X that Steiner requires.

The topic of Euclidean constructions suggests, then, that there are at least two kinds of explanatory proofs. First, we have what could be called pure explanatory proofs.[43] A pure proof is one that does not go outside the domain mentioned in the theorem. For example, a pure proof of a theorem about the natural

---

[42] An instance of this case is noted in Reutlinger et al. (2022).

[43] See Detlefsen and Arana (2011) for a discussion of purity.

numbers will involve only claims about the natural numbers. Second, we have what I will call abstract explanatory proofs. An abstract proof embeds the topic of the theorem in a different, more abstract domain. This is what we saw with the appeal to field extensions to make sense of Euclidean constructions. Pure and abstract explanatory proofs seem to have different virtues and vices. A pure proof clarifies the relations between entities in a given domain, but may for that very reason not clarify what is special about this domain as opposed to some other domain. One example of a pure proof is Euclid's bisection of the angle, where we learn not only that any angle can be bisected, but how to carry out the bisection in a transparent way. An abstract proof proceeds differently by factoring out many aspects of its target domain. This may lead one to lose some control or insight into the details of how that target domain works. For example, when we relate the possibility of a construction to a given field extension, it is not always clear how to carry out that construction. A famous instance of this loss of control is the constructability of regular n-sided polygons. Using the features of field extensions noted earlier, one can show that regular n-sided polygons can be constructed for $n = 17$, $n = 257$, and $n = 65,537$. However, this does not directly indicate how to construct such figures (Stewart 2004, chapter 19). So, while a great deal is gained by shifting from Euclidean constructions to field extensions, something is lost as well.[44]

4.1.2 Lange on Explanatory Proofs

Lange's work on noncausal explanation is the most detailed and innovative investigation of this topic. In his 2017 book and many papers Lange has assembled a large number of interesting cases drawn from scientific and mathematical practice (Lange 2017).[45] Lange's approach is also refreshingly pluralistic, as he does not try to force all explanations into a single structure (Lange 2017, p. 371). When it comes to pure mathematics, though, Lange considers only one kind of proof to be explanatory: "At least in many cases, what it means to ask for a proof that explains is to ask for a proof that exploits a certain kind of feature in the setup – the same kind of feature that is outstanding (i.e., salient) in the result" (Lange 2017, p. 255).[46] The notion of a setup is clearest for a theorem of the form "All F's are G" (Lange 2017, p. 232). Some feature of being a G may

---

[44] See also Colyvan et al. (2018) for a similar contrast motivated by a different case from abstract algebra.

[45] Lange (2017) incorporates most of Lange's earlier work on these topics. See Lange (2018a, 2018b, 2019, 2021a, 2022b) for newer discussions.

[46] I take Lange's official position to be that some explanatory proofs may explain in some other way, although he often talks as if they all work in the same way. It is also important to note that Lange discusses explanations in pure mathematics that are not proofs (Lange 2017, section 9.6; 2018b).

strike mathematicians as salient in the sense that it is puzzling or otherwise unexpected. One such feature that Lange emphasizes is symmetry. The explanatory proof then shows how being an F involves this symmetry. This will then explain how the setup is responsible for the salient feature of the result. Lange is somewhat flexible on how this aspect of being an F is cashed out. We may need to recast what being an F involves to uncover the respects in which it exhibits the feature in question. Part of the motivation for this proposal is that it respects the principle that an explanatory proof of a theorem must show that the theorem's truth is not a coincidence. When the symmetry of the result is the salient feature, we can indeed show that this aspect is not a coincidence by tracing that symmetry back to an aspect of the setup.[47]

In the following critical examination of Lange's proposal I am willing to grant that Lange has identified one kind of explanatory proof. However, there are other kinds of explanatory proof that work differently. In these other cases, I will argue, the explanation will show that the theorem's truth is not a coincidence, but will do this by indicating how a salient feature like symmetry emerges from the setup, even though the setup does not exhibit that very feature. This suggests that even when an explanatory proof traces a salient feature of the result back to the very same feature in the setup, the proof explains in virtue of the way this tracing out occurs, and not simply the tracing itself.

One case Lange notes that meets his tests involves Varignon's theorem. This theorem says that for any quadrilateral, if the midpoints of adjacent sides are connected, then a parallelogram results. Here the salient feature of the result is the symmetry of the parallelogram: its opposite sides are parallel. This is initially puzzling, as many quadrilaterals fail to exhibit any kind of symmetry, as with the figure ABCD in Figure 8.

Varignon's theorem can be proven using the "midpoint theorem" that says that "for any triangle, the segment connecting the midpoints of two sides is parallel to the third side" (Lange 2017, p. 254). For example, in Figure 9, triangle EFG has midpoint H on EF and I on FG. Line HI is parallel to the third side of the triangle EG.

To apply the midpoint theorem, first draw the diagonal AC in quadrilateral ABCD (Figure 10). Then the theorem tells us that the line connecting the midpoint of AB to the midpoint of CB is parallel to the diagonal AC. In addition, the line connecting the midpoint of AD to the midpoint of CD is parallel to the diagonal AC. So, these two midpoint lines are parallel and give us two of the opposite sides of our desired parallelogram. The other

---

[47] Other features that may be salient for Lange include unity and a contrast.

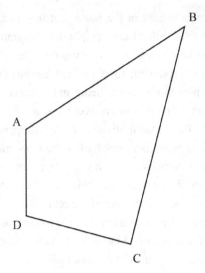

Figure 8 A quadrilateral.

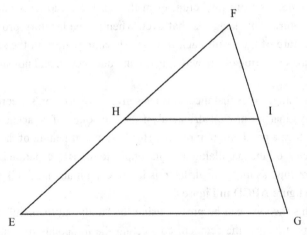

Figure 9 The midpoint theorem.

two sides can be shown to be parallel by a similar application of the midpoint theorem to the triangles involving the other diagonal BD. According to Lange, "the proof works by tracing the parallelism in the midpoint quadrilateral back to a parallelism in the original figure's triangles" (Lange 2017, p. 254). The explanatory proof involves exhibiting these triangles and illustrating how their symmetry is responsible for the symmetry of the resulting parallelogram.[48]

---

[48] For a classic textbook discussion of this proof, see Coexter and Greitzer (1967).

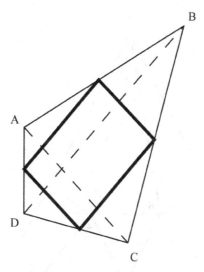

Figure 10 Varignon's theorem.

We can grant that some explanatory proofs work in this way. Still, can a proof explain a result, and thereby show that the result is not a coincidence, *without* tracing the salient feature of the result back to that very feature of the setup? It seems to me that the answer to this question is "yes." To motivate this answer I will discuss another case that Lange invokes right before he reviews Varignon's theorem. This is Morley's theorem (Lange 2017, pp. 252–253). This theorem says that for any triangle ABC, if we trisect each angle of the triangle using two lines, then these lines will cross within ABC to produce an equilateral triangle XYZ (see Figure 11).

Lange notes how this theorem is often called "Morley's Mystery" or "Morley's Miracle," even though it has been proved in many different ways. For Lange the mystery is tied to there being no explanatory proof. That is, taking Lange's account for granted, "no known proof of the theorem works by uncovering and exploiting a similar symmetry in the setup" (Lange 2017, p. 252). Unlike in the case of Varignon's theorem, there is no proof that traces the symmetry of the equilateral triangle back to something about the setup involving the arbitrary triangle that we started with. Here Lange also notes that "[p]roofs of the theorem often proceed 'backward': by starting with an equilateral triangle and generating the larger triangle from it" (Lange 2017, p. 253). One way to clarify what a backward proof is would be to say that it starts with a new claim that is entailed by the theorem. Suppose also that this new claim appears unrelated to the setup. As the new claim is not related to the setup, this proof will fail to trace the salient feature of the theorem back to that feature in the setup. And so, by Lange's account, it will not be an explanatory proof.

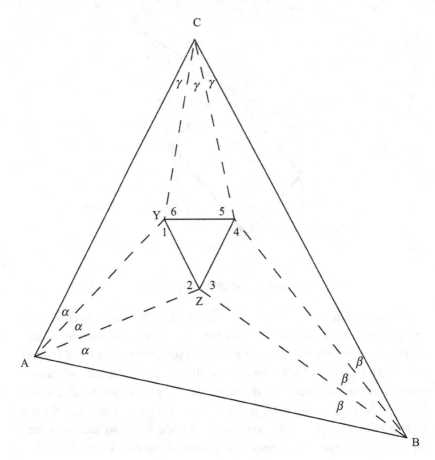

Figure 11 Morley's theorem.

However, there is a backwards proof of Morley's theorem that at least one mathematician deems explanatory. This is Conway's proof, which Conway characterizes as involving "finding a simple idea, so that any fool ... can understand the complicated thing" (Conway 2005, p. 1). More recently, Karamzadeh has used Conway's proof to claim that "the mystery of Morley's Theorem, is indeed, resolved" (Karamzadeh 2018, p. 298).[49] In broad strokes, Conway's proof involves using the angles α, β, and γ from Figure 11 to specify the angles of six new triangles. The proof then consists in showing how the lengths of the sides of these six new triangles can be found so that they fit together to divide up the area inside any arbitrary triangle so that an equilateral triangle results. In particular, Conway begins by stipulating that angles 1 and 4 are γ plus 60 degrees, angles 2 and 5 are β plus 60 degrees, and angles 3 and 6 are

---

[49] See Lange's methodological principle at (Lange 2017, p. 173).

α plus 60 degrees. This initially looks unexplanatory, as these six triangles seem to be unrelated to our starting point of an arbitrary triangle, and so it seems merely coincidental that they can be fit together as the theorem requires.

Conway's proof was criticized in something like these terms by Cain. In such cases the proof involves "an unexplained construction or a calculation of elements or a definition or a function that simply 'happens to work'" (Cain 2010, p. 8). When this happens, "[r]eaders cannot see why this construction or this calculation or that definition is being carried out; *they cannot perceive a reason for it that is internal to the proof*" (Cain 2010, p. 8, emphasis added). In Conway's proof, "[t]he values for what turn out to be the angles of the seven smaller triangles ... simply happen to work" (Cain 2010, p. 9).[50] Karamzadeh responds to these worries by supplementing Conway's original proof so that the choice of angles for his six new triangles is well-motivated: "Cain argues that Conway's specification of the angles is not justified retrospectively and that it's somehow pulled out of a hat. But, as I will show, they are in fact inevitable and quite explainable and therefore not a deus [ex machina]", namely not mysterious (Karamzadeh 2014, p. 4).[51] The key addition is the result that the internal triangle XYZ is equilateral just in case the angles fall into three equal pairs $1 = 4$, $2 = 5$, and $3 = 6$. So we can motivate Conway's choice of angles as "natural" (Karamzadeh 2014, p. 5) by showing that they are not only sufficient, but also necessary, for an equilateral triangle to be formed in this situation.

Once Karamzadeh's addition is made, we can grasp that the angles chosen by Conway for his six new triangles are, in fact, intimately related to the theorem's setup of an arbitrary triangle. Any triangle can be divided up into the triangles Conway describes, and conversely an equilateral triangle can be rescaled so that the six triangles Conway describes will combine to form whatever triangle one wishes. More generally, we can see how this sort of biconditional allows for explanatory proofs that target feature X of some theorem and do not involve the setup exhibiting feature X. Feature X can emerge from a setup that does not exhibit feature X, and yet still be explained.[52]

The same point holds for explanatory classifications like the Euclidean construction case discussed in Section 4.1.1. Here I suppose that the criterion articulated in terms of the degree of a field extension is explanatory because it

---

[50] Here Cain cites a newsgroup post by Conway from 1997 that presented the proof for the first time. This is the same proof as the one published in Conway (2005) and again in Conway (2014).

[51] See also Gorjian et al. (2015).

[52] Lange might respond that the symmetry of the equilateral triangle is traced back to the symmetry of the three equal pairs of angles. However, this seems to involve a different kind of symmetry than what we find in the equilateral triangle.

can be used to explain the possibility or impossibility of a given construction. However, as an abstract explanatory proof, this involves some loss of control over how to actually carry out a given construction. The proof provides a uniform criterion for possible constructions and yet does not provide any uniform method for carrying out a given construction. Similarly, there is a uniform criterion for impossible constructions, but this does not suffice to prove that a given construction is impossible.

This loss of control creates a problem for Lange: when he discusses this sort of case, he denies that an explanatory classification may treat the objects classified in different ways. This is because Lange supposes that the salient feature of an explanatory classification is the unity of the various parts of the classification. If unity is the salient feature, then to be explanatory "the proof must exploit some other, similar respect in which those cases are alike and must proceed from there to arrive at the result by treating all of the (classes of) cases in exactly the same way" (Lange 2017, p. 287). Consider, then, two possible constructions, for example bisecting an angle and constructing a square on a line segment, and two impossible constructions, for example trisecting an angle and squaring a circle. Bisecting an angle and constructing a square are unified by their possibility, while trisecting an angle and squaring a circle are unified by their impossibility. But our proof of the theorem involving degrees of field extensions does not afford a unified treatment of the constructions it classifies. We establish that the possible constructions are possible, but to show how the constructions work, additional specific steps need to be carried out. The same point holds for the other kind of construction: we can establish that the impossible constructions are impossible, but additional features of the points and their coordinates are needed to show this impossibility. For example, for trisecting the angle, one proof involves features of the cosine function. By contrast, to show that we cannot construct a square with the same area as a circle, we need to investigate the features of pi. In particular, if we show that pi is not an algebraic number (i.e. not the solution to any polynomial equation of any degree), then we have shown that no field extension that includes pi has the right degree (Stewart 2004, pp. 81–82). The upshot is that Lange is not able to endorse our classification of ruler-and-compass constructions as explanatory. He does not allow for explanatory classifications where there is a generic reason for some outcome across a range of cases, and yet that generic reason has to be supplemented with case-specific reasons.[53]

Lange does allow that explanatory proofs differ in degree in a way that is tied to how effectively the salient feature of the theorem is traced back to that feature

---

[53] Lange also suggests that explanatory classifications may have a contrast as the salient feature, but I do not think this avoids the basic worry I have raised. See Lange (2017, pp. 263–264, fn. 23).

in the setup. So, Lange might respond that the unity is still exploited here, but just to a lesser degree than in some other cases. It is not obvious, though, that these proofs are less explanatory in any sense, or less explanatory for this reason.

One central feature of Lange's proposal that we have not yet discussed is his contrast between natural and artificial properties. Lange says that natural properties involve "respects in which things may genuinely resemble each other" (Lange 2017, p. 335). A paradigm case of a nonnatural or artificial property is Goodman's predicate "grue": "being green and observed before the year 3000 [to update Goodman's example] or blue and unobserved before the year 3000" (Lange 2017, p. 336). We take green, but not grue, to be a natural property. Similarly, in mathematics, Lange argues that we take only some properties to be natural. Crucially, for Lange, only natural properties are apt to appear in explanatory proofs or the targets of explanatory proofs.

It is easy to see that if Lange failed to impose this restriction, then his whole approach to explanatory proof would count too many proofs as explanatory. To see why, return to the case of finite simple groups (Section 4.1.1). We can easily introduce a predicate "x is Gorensteinish" by defining it disjunctively in terms of the 29 kinds of finite simple groups. Then we would have the theorem that a finite group is Gorensteinish just in case it is simple. We could then trace back the salient unity of the finite simple groups back to their all being Gorensteinish.[54] However, given Lange's requirement that an explanatory proof involve only natural properties, he can rule out this case because it involves an artificial property.

The viability of Lange's proposal thus turns on this notion of a natural property.[55] Lange offers an interesting holistic proposal for characterizing natural properties that has both ontic and epistemic dimensions. First, on the ontic side, the naturalness of some properties and the explanatory character of some proofs arise together: "What makes a given proof ... explanatory is, in part, that it uses natural properties, and what makes those properties natural, in turn, is that they figure in other explanatory proofs" (Lange 2017, p. 338). On the epistemic side, one can use these holistic connections to discover a natural property or an explanatory proof: "Mathematicians discover that the properties in a given family are natural by finding them in many, diverse proofs that (mathematicians recognize) would be explanatory, if those properties were natural" (Lange 2017, p. 338). There is no reduction of being a natural property to something else, but only these robust connections that allow for such

---

[54] And, as Lange emphasizes, even stranger proofs could count as explanatory, for example a proof of the conjunction of two theorems from two unrelated domains (Lange 2017, p. 336).

[55] This is the sense in which everybody, including Lange, needs a restricted notion of mathematical ontology to make sense of explanatory proof. See the discussion of Povich in Section 3.2.3.

properties to be (defeasibly) identified. For example, being a group is a property that was independently identified in various mathematical domains. This motivated mathematicians to take being a group as a natural property, which in turn generated the desire to identify and explain the range of finite simple groups.[56]

To conclude our discussion of Lange's account of explanatory proofs, I would like to suggest that once we have natural properties, we can see that the natural properties that appear in an explanatory proof need not be as closely related to the target natural property as Lange supposes. In Conway's proof of Morley's theorem, as clarified by Karamzadeh, we have natural properties that explain the appearance of the equilateral triangle, but in a way that fails to fit Lange's account. In the explanatory classification of ruler-and-compass constructions, Lange's account also proved too stringent. So, it looks like we should adopt an account that is broader than Lange's account: a proof is explanatory just in case its target is characterized in terms of natural properties and the proof turns on natural properties. We can preserve Lange's holism and say that natural properties are just those properties that appear in many explanatory proofs. We have no reductive, independent characterization of either notion, but we can use this amended proposal to make sense of the wide variety of proofs that are counted as explanatory in mathematical practice. Of course, this does not settle what it means for a proof to turn on natural properties, so there is a residual puzzle for this proposal. I consider the options for dealing with this puzzle in Section 4.3.

4.2 Pluralist Accounts of Genuine Mathematical Explanations

4.2.1 Steiner on Genuine Mathematical Explanations

In a companion paper to the one discussed in Section 4.1.1, Steiner extends his account of explanatory proofs to an account of what we are calling genuine mathematical explanations (Steiner 1978b). Recall from Section 2.1 that these were initially identified as scientific explanations that involve mathematics, but where the mathematics is not functioning merely to represent some nonmathematical explainers like causes. Steiner proposes a simple test for genuine mathematical explanation: "when we remove the physics, we remain with a mathematical explanation – of a mathematical truth!" (Steiner 1978b, p. 19). This can be seen as an ancestor of Baron's 2019 thought that we should consider the mathematics that is not used to represent causes or other physical explainers (see again Section 2.1): we have explanatory information and yet "that aspect is not a relatum in a mapping relation between the mathematical and physical structures at issue" (Baron 2019, p. 710). Then that explanatory

---

[56] Wussing (2007) is a classic discussion of this history.

information may extend to associated physical structures, but not simply because the mathematics is representing causes.

In his 2012 article "Science-Driven Mathematical Explanation," Baker convincingly argues that Steiner's account of genuine mathematical explanations does not work (Baker 2012). Baker invokes genuine mathematical explanations where no proof is judged to explain the relevant theorem. Baker's main case here is the fact that bees use hexagons to build their honeycombs. This is explained using the theorem that hexagons are the most efficient way to tile a two-dimensional plane while minimizing the tiling materials. While this explanation seems to be a good one, and also one that qualifies as a genuine mathematical explanation, there are no explanatory proofs of the theorem employed. The same point can made with our example of the bridges of Königsberg: we showed how this amounted to a genuine explanation of the impossibility of the bridge crossings, and yet no reference was made to the kind of proof that the mathematical theorem has.

One could also object to Steiner based on nonexplanations that satisfy all of his conditions. Consider an uncontroversial case of an explanatory proof from pure mathematics like the proof of Varignon's theorem sketched in Section 4.1.2. The theorem says that for any quadrilateral, if the midpoints of adjacent sides are connected, then a parallelogram results. Consider any explanatory target that has the same form as the theorem such as that all metal wires, when connected to a battery, conduct electricity. Then we can interpret "quadrilateral" as standing for being a metal wire, "connecting adjacent midpoints" as standing for being connected to a battery, and "resulting in a parallelogram" as standing for conducting electricity. This interpretation of the mathematical theorem then transforms the proof of the theorem into a derivation of the claim that all metal wires, when connected to a battery, conduct electricity. But nobody would count this as an explanation. Still, Steiner's test is met: "when we remove the physics, we remain with a mathematical explanation – of a mathematical truth." So it should be clear that this test is not a good one. One response to this objection is to require a more demanding kind of embedding than the simple reinterpretation that I have sketched here. However, it is not clear how to rule out this sort of problematic case unless these embeddings are characterized in explanatory terms.

4.2.2 Lange on Genuine Mathematical Explanations

In line with his pluralism, Lange allows for a variety of kinds of genuine mathematical explanations or, as he puts it, "distinctively" mathematical explanations. The most important kind of genuine explanation for Lange is

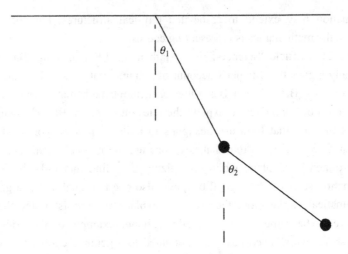

Figure 12 A double pendulum.

what he calls an "explanation by constraint."[57] Lange starts by supposing that causal explanations work through the use of causal laws, and that causal laws have their own modal strength, namely physical necessity. Consider a case where the explanatory target has a modal strength that is greater than physical necessity. Many suppose that mathematical truths have a special sort of necessity that is greater than physical necessity. This means that if a mathematical truth can be appropriately related to some physical system, then the mathematics may explain why that physical system exhibits some features with a modal strength that is greater than physical necessity. This provides what Lange calls a type-(m) explanation by constraint (Lange 2017, p. 131).[58]

One of Lange's examples involves a double pendulum (Lange 2017, pp. 26–29). This is a physical system where one pendulum is attached to another, as in Figure 12. There are at least four equilibrium configurations for any double pendulum: θ_1 is either 0 degrees or 180 degrees, and θ_2 is either 0 degree or 180 degrees. Lange shows how in this case there are two kinds of derivations of the number of equilibrium configurations. Both derivations appeal to the system's potential energy U. One derivation develops an equation for U as a function of the angles α [θ_1] and β [θ_2] by considering the laws that govern the forces at work in such a system in the actual world, for example the force of gravity. However, Lange outlines another derivation that "exploits merely the fact that by virtue of the

---

[57] See especially Lange (2017, chapters 1–4). Other sorts are really statistical explanations (chapter 5) and dimensional explanations (chapter 6) that I must set aside for reasons of space.

[58] Other types are type-(c), whose targets are themselves constraints, and type-(n), whose targets arise in part due to contingent facts (Lange 2017, p. 131).

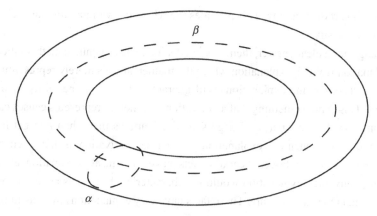

Figure 13 A torus.

system's being a double pendulum, its configuration space is the surface of a *torus* – that is, that U is a function of α and β" (Lange 2017, p. 27, emphasis added). This happens because, for any angle, that angle plus 360 degrees is the very same angle. So we get a circle of potential angles for α and another circle of potential angles for β. These possibilities for $U(\alpha, \beta)$ can be conveniently associated with the points on the surface of a torus or doughnut, as in Figure 13.

Once this association is carried out, the character of the torus ensures that no matter how $U(\alpha, \beta)$ is fixed, namely no matter the force laws at work, there will be at least four equilibrium configurations. This is a matter of the topology of the torus. For Lange, the second derivation that draws on the features of a torus and relates it to the double pendulum is an example of a genuine mathematical explanation. This is because it explains why the double pendulum has at least four equilibrium configurations in a way that shows that this number is more than physically necessary.

In general, for Lange, "these explanations work not by describing the world's network of causal relations in particular, but rather by describing the framework that any physical system (whether or not it figures in causal relations) *must* inhabit, where this variety of necessity is stronger than the necessity possessed by ordinary laws of nature" (Lange 2017, p. 44). This does not mean that the explanation is completely mathematical. In the double pendulum case, we must suppose that equilibrium configurations can be identified with the so-called stationary points of $U(\alpha, \beta)$, the potential energy function. But Lange argues that this assumption uses only Newton's second law, $F = ma$, and that this law has greater necessity than an ordinary law of nature like Coulomb's law or the law of universal gravitation. If we grant this point, then we can see how every step in the derivation is either purely mathematical or else involves a law with a special sort of necessity. This provides the whole derivation with that special sort of necessity.

Lange has a clear answer, then, to what distinguishes a genuine mathematical explanation from an explanation where the mathematics is merely representing causes. The genuine explanations will guarantee the special necessity of the target. This is not something that a derivation that merely represents causes can accomplish. For this reason, Lange sometimes argues that when a target has a genuine mathematical explanation, the causal derivation is not even an explanation of that target: "certain facts have no causal explanation at all because any such explanation would mischaracterize these facts as more contingent than they actually are. These facts have only explanations by constraint" (Lange 2017, p. 47).[59]

Lange extends his account of genuine mathematical explanations to many cases, including the bridges case that we introduced in Section 2. To handle many of these cases, Lange must treat the target as including some contingent matters of fact. There are thus two ways to obtain a genuine mathematical explanation: "the facts doing the explaining are eligible to explain [either] by virtue of being modally more necessary than ordinary laws of nature (as both mathematical facts and Newton's second law are) or being understood in the why question's context as constitutive of the physical task or arrangement at issue" (Lange 2017, p. 33). In the bridges case, when we ask why everyone has failed to complete a certain tour of the bridges, Lange supposes that we understand the actual arrangement of the bridges to be constitutive of the task. Of course, this actual arrangement is highly contingent. But its contingency is discounted. The target is then that, given that the bridges have their actual arrangement, why have the attempts to complete these circuits failed? The answer establishes the necessity of this outcome in a way that no causal explanation could.

Lange's move to allow some contingent facts to be regarded as constitutive has generated some concerns that his proposal counts too many derivations as explanatory. One objection is that every target that may be described in mathematical terms is amenable to some genuine mathematical explanation (Pincock 2015a, p. 875; Craver and Povich 2017, pp. 35–36). Consider, for example, the flagpole case. This involves a mathematical fact, some contingent features of the situation, and a physical law. So, if we regard the contingent features and law as constitutive of the situation, then the length of the shadow will be subject to

---

[59] Lange does make clear that some genuine mathematical explanations are what he calls "hybrid explanations" that combine constraints that are more than physically necessary and causal laws that are merely physically necessary. See Lange (2017, section 2.3). But I set those cases aside here, again for reasons of space.

a genuine mathematical explanation. In this reply to Craver and Povich, Lange seems willing to accept this point (Lange 2018a, p. 87, citing Lange 2017, p. 134, fn. 35): the same target, namely the length of the shadow, may or may not have a genuine mathematical explanation, depending on what we regard as constitutive of that situation, for example the actual height of the flagpole and position of the sun.

A related objection to Lange's proposal is that it is vulnerable to pairs of derivations that violate explanatory priority. Recall that this is the principle that if A is part of an explanation of B, then B is not part of an explanation of A. Consider any biconditional that is mathematically necessary, such as the theorem deployed in the bridges case: a graph has an Euler path just in case exactly zero or two nodes have an odd valence. The bridges of Königsberg have nodes with five, three, three, and three edges, and so (A) they do not include exactly zero or two nodes with an odd valence, and so (B) they do not contain an Euler path. Conversely, the bridges of Königsberg (B) do not contain an Euler path, and so (A) they do not include exactly zero or two nodes with an odd valence. When combined with the mathematical theorem, A is part of a derivation of B, and B is part of a derivation of A. All of Lange's modal conditions are met, so Lange's proposal commits him to counting both as explanations. This is a violation of explanatory priority. This is all the more troubling, as Lange has endorsed a version of explanatory priority.[60]

One response that Lange has to this objection is that we do not, in fact, have a violation of priority. For the explanation of B must treat certain matters as constitutive of the situation, and the explanation of A must treat other matters as constitutive of the situation. So, we only have A being part of an explanation of B', and B being part of an explanation of A', where the primes indicate that different matters are being treated as constitutive. The apparent violations of priority are then simply a further manifestation of the point made earlier that any target can receive a genuine mathematical explanation if the right elements are treated as constitutive.

Another response that Lange has to priority objections is that there is a principled barrier to regarding something as constitutive of the explanatory target. Lange makes this sort of point in his book, and also emphasizes it in his reply to Craver and Povich (Lange 2018a). Craver and Povich object to this aspect of Lange's view using their directionality requirement (Section 3.2.3). I raise essentially the same worry using our test for explanatory priority. Consider a city L that, unlike Königsberg, has a network structure Q that permits

---

[60] See Lange (2009, p. 207): "Relations of explanatory priority are asymmetric." Some qualifications to this principle seem to be introduced at Lange (2017, p. 298, fn. 20), in line with the suggestion I develop in the next paragraph.

an Euler crossing (as in Figure 7), and the proposed explanation that L's bridges have that structure because they have been crossed. Lange insists that "having actually been crossed is no part of what it is for an arrangement of bridges to be the arrangement at issue" in the explanatory question. This is because the question "is understood to concern an arrangement of bridges that would still have been that arrangement even if no one ever crossed it" (Lange 2017, pp. 42–43; noted at Lange 2018a, p. 86). Here Lange is saying that it is not up to us to regard this or that feature as constitutive. There is instead a modal test that must be passed for something *to be* constitutive: if being crossed is constitutive of these bridges having this structure, then if no one had ever crossed the bridges, the bridges would not have had this structure. So, if a scientist makes a mistake and regards something as constitutive when it is really not, then we do not have a legitimate explanation. By generalizing this proposal, Lange may hope to avoid all violations of directionality and even priority.

It is not clear to me that this proposal works for the violation of priority sketched earlier. For Lange to get the explanation that he counts as legitimate, it must be appropriate to regard the actual network structure of the bridges of Königsberg as constitutive. That makes it legitimate to appeal to this structure to explain the absence of the Euler path. To avoid the violation of priority, Lange could say that the absence of the Euler path is not constitutive of this system. However, the absence of the Euler path is necessary and sufficient for the bridges to not include exactly zero or two nodes of even valence, and this biconditional obtains by a strong sort of mathematical necessity. So, no modal test of the sort sketched in the previous paragraph can discriminate these conditions: in all possible worlds, if one of these properties is constitutive, then so is the other. I conclude that violations of priority remain even if modal restrictions are placed on constitutive properties.[61]

The last sort of objection to Lange that I will consider focuses on modality. Is the modality of some target really so intimately connected to genuine mathematical explanation? In the pure mathematics cases, we saw in Section 4.1.2 how Lange does not analyze explanatory proofs in modal terms. This may be because all mathematical truths are thought to be modally on a par. This shows that some explanations establish relations among truths that are not modally distinguished from one another. Perhaps the same relations can obtain between mathematical truths and contingent facts. To appreciate the issue, consider again Lange's double pendulum case. Newton's second law and some mathematical truths ensure that a double pendulum has at least four

---

[61] Only something like an essence that is individuated in hyperintensional terms could avoid this worry. This is consistent with Lange's discussion in Lange (2021b), which I briefly discuss in Section 5.2.

equilibrium configurations. If a scientist asks why it is necessary that the system has at least four equilibrium configurations, then the modality is incorporated into the target, and so arguably a legitimate explanation must indicate why this state of affairs is necessary. So, for this kind of why question Lange's emphasis on modality is very plausible, as the modality is explicit in the why question.[62] However, the situation is not as clear for a scientist who merely asks why the system has at least four equilibrium configurations. Lange says that this why question must also receive an answer in terms of truths with the right kind of necessity. Otherwise, the proposed answer "would mischaracterize these facts as more contingent than they actually are" (Lange 2017, p. 44). But if the modality of the target is not at issue, it is not clear what sort of mischaracterization this is.

One clarification of Lange's position would be to say that a nonmodal explanation is illegitimate because it presents a false picture of what the target depends on: in fact, the target depends on something necessary, so to explain the target by appeal to something else is just to say something false. Similarly, to say that no has crossed the bridges of Königsberg exactly once on a tour because they were just built is to fail to explain the target. To develop this sort of response, Lange would have to clarify how the modal explanation of the modal target excludes the nonmodal explanation of the nonmodal target. The view I am suggesting is that both explanations are legitimate and both present accurate accounts of what their respective targets depend on. If this is right, then we could have genuine mathematical explanations that do not explain in virtue of their mathematical necessity.

This point is even more plausible when we consider a case like the bridges case where contingencies abound. One sort of explanatory target will include the modality in target, as with the question "Why is it impossible for anyone to carry out this sort of tour of the bridges of Königsberg?" It is not absolutely or metaphysically impossible, as many changes to the bridges would allow the tour. But it is plausible that a legitimate explanation for this target will account for this impossibility, in part by clarifying its strength. This can involve regarding various features of the system as constitutive, as Lange emphasizes. However, I do not see why this implies that an explanation of a nonmodal explanatory target should proceed in the same way. For example, if the question is "Why have all these people failed to carry out this sort of tour of the bridges of Königsberg?," then an adequate answer is simply that there is no such path. The modality of this absence does not need to be specified if it is not part of the explanatory target.

---

[62] This is an (m)-type case for Lange (2017, p. 131).

On this reading, Lange has transferred a feature of some genuine mathematical explanations (his (m)-type) to a feature of all genuine mathematical explanations (including his (n)-type). When the target has a strong form of necessity included in it, then we get explanations that work as Lange describes. However, in other cases, where the target does not have that strong form of necessity included in it, we can have other forms of genuine mathematical explanation. Lange may object that these other forms have not been suitably circumscribed: in line with our first principle for accounts of explanation, what is the difference between a mere mathematical description and a genuine mathematical explanation, if we drop the modal tests? We can ask Lange the very same question when it comes to explanatory proofs. Here, there are no modal tests. Lange's account requires that the proof turn on natural properties and that the proof trace the salient feature of the result back to the setup. This also does not clarify how this "tracing out" must occur in order for the proof to count as explanatory.

An answer to both of these questions about explanation is that there is no way to identify the difference in more basic terms. For proofs, there is a difference between explanatory and nonexplanatory proofs, but we cannot break that difference down to anything more basic. For genuine mathematical explanations, there is also one or more basic notions of explanatory relevance that link mathematical facts to nonmathematical facts, but we cannot provide any reductive analysis of what these relations come to. Instead, we can enumerate the various relations that we find arrayed across these causal and noncausal cases. This sort of brute pluralism is sometimes defended in terms of an irreducible variety of dependence relations. For example, Koslicki maintains that "an explanation, when successful, captures or represents ... an underlying real-world relation of dependence of some sort which obtains among the phenomena cited in the explanation in question" (Koslicki 2012, pp. 212–213; noted at Pincock 2015a, p. 878). The brute pluralist is committed to there being more than one type of dependence relation. But they are not committed to there being any reductive analysis of what these dependence relations amount to. Perhaps this brute pluralism is the best account of genuine mathematical explanation and explanatory proof.

4.3 Pluralism and the Value of Explanation

Pluralists about explanation face a problem that monists can avoid: if different kinds of explanation work differently, then what do all explanations have in common that make them count as explanations? The brute pluralism sketched at the end of the last section has difficulty giving much of an answer to this question. On the one hand, the brute pluralist is a pluralist, and so they must insist that the different sorts of explanatory relevance relations are different. A causal explanation will turn on causal relevance, and the different sorts of noncausal explanations will have special relevance relations that distinguish them from causal explanations. One model for this sort of relationship is a genus/species model: objects of various species can be grouped under a common genus. For example, all regular polygons have an illuminating common feature, but squares are still different from pentagons in a number of identifiable respects. In other sorts of cases with a genus/species structure there is not much more to say: some properties are determinates of a common determinable, and yet the contrasts among determinates are basic. For example, one shade of blue is different from another shade of blue in an elusive and primitive way. Similarly, the brute pluralist may insist that one kind of explanation is different from another kind of explanation in a primitive way. We can perhaps grasp these differences and also see what all cases have in common, but no further elucidation is possible.

As we have seen, Lange has a more ambitious approach that aims to say something more informative about explanation. In the final chapter of his book, Lange argues that all varieties of explanation are "species of the same genus" (Lange 2017, p. 399). This argument turns on the answer to the following question: how do natural properties figure into explanations by constraint? More precisely, Lange asks, for a law L and property P, "When is P natural enough for its instantiation to be eligible to join with L and other initial conditions in explaining the outcome?" (Lange 2017, p. 386). The background assumption here is the same as we saw in Section 4.1.2: explanations require natural properties. However, in the scientific case Lange does not endorse the holist proposal that being an explanatory proof and including natural properties arise together. Instead, Lange's view is that when P is a nonfundamental physical property (Lange 2017, p. 371), "P is natural enough to explain if and only if there is an explanation of L in which P (that is to say, the combination of properties to which it reduces) enters as a unit" (Lange 2017, p. 386). So, the availability of a property to explain in scientific contexts is fixed by how laws may be explained. These explanations of laws may be ordinary causal explanations or the more involved explanations by constraint, including genuine mathematical explanations, that we discussed in Section 4.2.2.

The links between natural properties and explanation in both mathematics and science highlight how Lange's pluralism is different from brute pluralism. Lange here argues that there are features that all explanations have in common, despite their differences. The most important common feature is that all explanations "are alike in their capacity to render similarities non-coincidental" (Lange 2017, p. 398). This claim relies on substantial links between explanation, natural properties, and similarities. For Lange, sharing a natural property just is being similar in some respect. So when explanations proceed via natural properties, they involve similarities. In the special case where the salient feature of a target is itself some similarity that obtains in some domain, the explanation will account for that similarity and thus render it noncoincidental. More generally, for any salient feature, the use of natural properties in the explanation functions to remove the appearance that the feature is coincidental. As explanations all share at least the common feature of removing apparent coincidences, Lange concludes that "the property of being an explanation is a natural property, too" (Lange 2017, p. 399).

Another common feature of explanation, for Lange, is tied to the value of explanation. According to Lange, explanations have a distinctive, intrinsic value: "Like science, mathematics values explanatory power as an end in itself, not merely as a means for discovering techniques that might prove as yet unproved theorems (or suggest new theorems) and not as a means for finding proofs that exhibit other virtues" (Lange 2017, p. 292). Lange does not clarify how the natural property of removing apparent coincidences is related to this intrinsic value.

In arguing for substantial common features that unite explanations, Lange's pluralism about explanation marks an improvement on a brute pluralism that simply lists all the kinds of explanations that there are. Notice, though, that Lange's arguments that explanations have these common features does not draw on the aspects of his account of explanation that we have called into question in Sections 4.1.2 and 4.2.2. All we need is a notion of a natural property and the principle that explanations and explanatory targets involve natural properties. We do not even need to suppose, as Lange does for explanatory proofs, that natural properties arise through their role in explanatory proofs. There may just be an open-ended list of explanatory relevance relations (or dependencies) that relate explanations and targets that involve only natural properties. But these natural properties could arise for any number of reasons that are unconnected to their role in explanation. Of course, this might reintroduce the sort of brute emergence that Lange is eager to avoid, but it is not clear why these aspects of Lange's proposal count in its favor.[63]

One common feature that Lange emphasizes is the value of explanations for mathematicians and scientists: they strive to get these explanations. The efforts expended to find explanations, despite the difficulties, show that they are important. However, this does not require that we agree with Lange and

---

[63] See also Rice and Rohwer (2021) for the proposal that explanations merely exhibit a family resemblance to one another.

maintain that the value is intrinsic. Other views of the value of explanation would emphasize their cognitive or epistemic role. Lange seems to endorse a derivative value for explanation as well as its intrinsic value: a cognitive value for explanation is that it removes apparent coincidences. Suppose that coincidences were somehow disturbing or undesirable. Then explanations would have a derivative cognitive value due to their capacity to render coincidences merely apparent.

Another proposal is that explanations are epistemically valuable due to their role in increasing our knowledge. This proposal is sometimes developed by defenders of what is called "inference to the best explanation" (IBE). According to IBE, in certain situations, a proposed explanation's being the best of the available explanations is some evidence that the proposed explanation is a legitimate explanation. So an agent who carries out this inference can come to justifiably believe some claims based on this special explanatory evidence. In cases where this explanation is legitimate, and so these claims are true, this agent may even come to know something new. As scientists and mathematicians value knowledge, the way that explanations can increase our knowledge makes explanations derivatively valuable.

4.4 Conclusion

This section has considered two pluralist proposals for explanatory proofs and genuine mathematical explanations. Steiner's work was innovative, but ultimately limited by the tension between his proposals and the cases we considered. Lange's account is considerably more developed and plausible. We saw no reason to deny that some cases work in the way that Lange claims. However, I argued that Lange's account is not pluralist enough, as there are cases that seem to elude his framework. The upshot is an even more pluralist proposal that posits additional explanatory relevance relations that do not involve modality. While this brute pluralism is philosophically disappointing, it is consistent with the genus/species model that Lange develops. It can also be motivated by the assumption that explanations have distinctive cognitive and epistemic roles in both mathematics and science.

5 Autonomy, Indispensability, and Inference to the Best Explanation

Clarifying how mathematical explanations work is an important part of clarifying how the subject matter of mathematics relates to the subject matter of sciences like physics, psychology, and economics. In this section I consider how the relationship between mathematics and physics is both similar to and different from the relationship between psychology and physics. To start, I will argue that mathematics is autonomous from physics by comparing mathematical explanation to psychological explanation (Section 5.1). This is followed by an examination of some influential arguments for a platonist interpretation of mathematics (Section 5.2). A platonist claims that mathematics is about a special domain of abstract objects. An "indispensability" argument for platonism relies on mathematical explanations having a certain character. I will try to identify what this character is and then turn to a brief consideration of how an appeal to inference to the best explanation (IBE) argument could deliver a platonist interpretation (Section 5.3).

5.1 Putnam on the Autonomy of the Special Sciences

What sort of autonomy should we claim for pure mathematics? One classic discussion was offered by Putnam. His "Philosophy and Our Mental Life" argues for a significant analogy between how psychology relates to physics and how mathematics relates to physics. In the case of both psychology and mathematics, unlike physics, what is important is that a theory provide a description of a range of systems that are related by a "functional isomorphism" (Putnam 1975, p. 292). A functional isomorphism obtains when there is a function that maps the objects, properties, and relations of one system to the objects, properties, and relations of the other system in such a way that the structure of the first system is preserved in the second system. Putnam develops this point for mathematics using the example of a cubical peg and a board with two holes where "the peg passes through the square hole, and it does not pass through the round hole" (Putnam 1975, p. 295, emphasis removed). Putnam argues that an explanation of this fact involves the geometrical features of the peg and the hole: the cubical peg has a height of 15/16″, the circular hole is 1″ in diameter, and the square has sides that are 1″ long. For Putnam, "The explanation is that the board is rigid, the peg is rigid, and as a matter of geometrical fact, the round hole is smaller than the peg, the square hole is bigger than the cross-section of the peg" (Putnam 1975, p. 296). For the peg to fit through the round hole, the diagonal line connecting two corners of some side of the peg must be less than the diameter of the hole. But each of these diagonal lines is about 1.33″ in length (via the Pythagorean theorem). So this is why the peg fails to go through the round hole. But the square hole is larger than this side of the square peg, and so it may pass through.

Putnam calls the geometrical explanation of this fact "autonomous": "The same explanation will go in any world (whatever the microstructure) in which those *higher-level structural features* are present. In that sense *this explanation is autonomous*" (Putnam 1975, p. 296). Crucially, a mathematical description will pick out such structural features, and it is these very features that will be preserved in the functional isomorphisms between systems that the description affords. This makes mathematics into an autonomous science in the sense that it provides autonomous explanations. Autonomous sciences give explanations that are different in kind from sciences whose explanations require an appeal to the microstructure of some target system. A microstructural explanation would not be preserved across systems with different microstructures. Putnam is happy to admit that physics often provides microstructural descriptions that entail that some fact will obtain. For example, in this case Putnam allows that there is a microstructural physical description of the board and peg that entails that the peg may pass through the square hole and may not pass through the round hole. But he questions whether or not this deduction is an explanation: "What makes you call this deduction an explanation?" (Putnam 1975, p. 296). Later on the same page, he simply calls this sort of deduction a terrible explanation (Putnam 1975, p. 296).

Putnam's discussion suggests two versions of autonomy. The *strong* autonomy of some science like geometry or psychology requires that this science provides explanations of some phenomenon, and physics cannot explain this phenomenon. By contrast, *weak* autonomy requires only that this science provide an explanation of some phenomenon that is better in some respect than an explanation that physics could provide. Either form of autonomy can be motivated by focusing on what Putnam calls "higher-level structural features" such as shapes and functional isomorphisms at this level of description. If such features turn out be explanatorily relevant to some phenomenon, then a science that identifies these features and their relationships to one another has a potential explanatory advantage over physics. Of course, this advantage turns on the assumption that physics is not well suited to identify or study these higher-level structural features.

The main aim of Putnam's paper is to argue that we know enough about the mind to conclude that any adequate psychological theory will focus on some higher-level structural features. As a result, we can be assured that psychology is autonomous with respect to physics. Our focus is on the parallel question for mathematics: should we say that the mathematical explanations found in science provide evidence for the autonomy of mathematics with respect to physics? This question splits into two versions based on the contrast between strong and weak autonomy. I will argue here that strong autonomy is not plausible, but that weak autonomy is well supported.

Strong autonomy requires that there are mathematical explanations for some phenomena and no physical explanations for at least one of those phenomena.

One objection to strong autonomy is based on the supervenience principle that we introduced back in Section 3.2.2: there is some basic physical level for any system such that a change in any feature of the system will go along with a change in that basic physical level of the system. Suppose one claimed that the mathematical explanation of the peg/board pattern is unexplainable by physics. That is, there is no possibility of completing our physics in a way that would explain why the square peg passes through the square hole but not through the round hole. If we accept the supervenience principle, then it seems clear that this pattern is explainable by physics. For consider the salient higher-level structural property that we appealed to in explaining why the square peg could not pass through the round hole. This property is that the length of the diagonal of one side of the peg is around 1.33″, which is greater than the 1″ diameter of the round hole. If supervenience holds, then there is some purely physical description of a feature F of the peg that varies with the length of the diagonal: if the length were different, then F would be different as well. So, it seems that any explanatory job that the length is doing could be performed by a purely physical description of F, for example supporting a causal counterfactual. Once we extend this point to all of the mathematical properties and relations that appear in the mathematical explanation, we can conclude that there must be, at least in principle, a physical explanation for this peg/board pattern. Furthermore, we might insist that this physical explanation is better in some respects than the mathematical explanation. For the physical explanation can be expanded to account for features of the peg and board that the mathematical explanation took for granted. For example, the mathematical explanation simply assumes that the peg and board are rigid, while the physical explanation could include aspects of the microstructure that are responsible for this rigidity.

Note that the existence of these physical explanations does not entail that mathematics is reducible to physics. Such a reduction requires not only the physical explanations, but also "bridge laws" that link each mathematical property to physical properties. As a mathematical property like being a cube is arguably realized by a heterogeneous variety of physical properties, these linking bridge laws will not have the right character for a reduction. The key point for my argument against strong autonomy is not the existence of a reduction, but just the existence of a physical explanation for each target of a mathematical explanation.

These concerns about strong autonomy do not undermine weak autonomy. Putnam in particular emphasizes both the generality and the robustness of the mathematical explanation of the peg/board pattern. There is good reason to think that the physical explanations required by supervenience will not typically exhibit these explanatory virtues. So we can conclude that some mathematical explanations of some phenomena will be better in these respects than the physical explanations of those phenomena. Consider first generality. The mathematical explanation of the peg/board pattern deploys the properties of lengths and rigidity in place of the more basic physical properties that hold for that particular system. As other systems with other microstructures may exhibit the same lengths and rigidity, the mathematical explanation applies more generally than the microstructural physical explanation. These points are closely related to a second explanatory virtue that I will call robustness. In Putnam's terms, attributed to Garfinkel, "a good explanation is invariant under small perturbations of the assumptions" (Putnam 1975, p. 301, emphasis removed). Small changes to the lengths in question or rigidity will not stop the mathematical explanation from going through. That is, the mathematical explanation of what is going on in that one system is superior to an explanation that is fragile in the sense that it fails to apply under small changes. But there is good reason to think that the microstructural physical explanation is fragile in just this sense. For some small variations in the microstructure will destroy the rigidity of the peg or board, and thus alter the situation by, for example, producing a crack in the peg. I conclude that mathematical explanations that target higher-level structural features tend to exhibit the virtues of generality and robustness that the corresponding physical explanations of those targets tend to lack. That is, there is good reason to endorse the weak autonomy of mathematics with respect to physics.[64]

5.2 Explanatory Indispensability Arguments for Mathematical Platonism

Many recent discussions of mathematical explanation are motivated by what are called "enhanced" or explanatory indispensability arguments for platonism. A platonist says that mathematics is about a special domain of abstract objects. These objects are outside of space and time and so are causally isolated from concrete, physical objects. Colyvan and Baker can be credited with advancing these new indispensability arguments in their strongest form (Colyvan 2001, 2002, 2010, 2012; Baker 2005, 2009b, 2012, 2021). In his 2001 book *The Indispensability of Mathematics*, Colyvan presented a general indispensability argument that he traced back to Quine and Putnam (Colyvan 2001, p. 11). This argument was then refined by Baker to focus more directly on explanatory considerations:

1. We ought rationally to believe in the existence of any entity that plays an indispensable explanatory role in our best scientific theories.

---

[64] There is a vast literature on these arguments that tends to focus on psychology. See Lyon (2012) for an approach to mathematical explanations that uses Jackson and Pettit (1990) to develop an account of mathematical explanations in these terms.

2. Mathematical objects play an indispensable explanatory role in science.
3. Hence, we ought rationally to believe in the existence of mathematical objects (Baker 2009b, p. 613).

The entities that an agent ought to believe in are sometimes termed their "ontological commitments." In Quine's influential formulation, encoded in premise 1, the entities that one takes to exist should be identified through a regimentation of our best scientific theories. In the course of this regimentation, many apparent ontological commitments will be shown to be not genuine, as with the average household or shadows. But the remaining ontological commitments that cannot be eliminated or dispensed with are the ones the agent is stuck with. Otherwise, in Putnam's apt formulation, one is embracing "the intellectual dishonesty of denying the existence of what one daily presupposes" (Putnam 2010, p. 57).[65]

Baker adds a new emphasis on mathematical explanation to this traditional account of ontological commitment. He can allow the existence of nonmathematical versions of our best scientific theories. To show that premise 2 is true, one need only argue that these nonmathematical versions afford either no explanation of what the mathematized theory can explain, or else only worse explanations of what the mathematized theory can explain. Either way, we have sufficient reason to include mathematical entities in our ontological commitments. The debate has thus been conducted by focusing on alleged cases of mathematical explanations of physical phenomena and the potential for nonmathematical explanations of those phenomena.

Our discussion so far has suggested two different ways of motivating the explanatory indispensability of mathematical entities. In Section 5.1 I sketched the argument that mathematics is weakly autonomous with respect to physics. The weak autonomy of mathematics requires the existence of mathematical explanations that are better in some respects than any explanation afforded by physics, for example in terms of generality or robustness. The other approach, surveyed in Sections 2–4, is to identify a special class of genuine mathematical explanations, where this genuineness is provisionally characterized in terms of the mathematics *not* explaining by merely representing some physical causes. Our question, then, is whether either of these approaches provides adequate support for premise 2.

The appeal to weak autonomy has the virtue of being very plausible, but the vice of failing to make appropriate links to the existence of mathematical entities. A review of Section 5.1 shows that we made the case for the weak autonomy of mathematics with respect to physics without taking a stand on the existence or

---

[65] A different motivation for premise 1 is considered in Section 5.3.

character of any mathematical entities. This suggests that it may be possible to endorse the mathematical explanations provided by our best scientific theories without including any mathematical entities in our ontological commitments. This seems to be not only a possible attitude, but also the right attitude for cases like Putnam's peg/board explanation. For as Hartry Field famously argued, these sorts of explanations involve an extrinsic relation between the peg/board system, some real numbers, and some standard unit, for example the inch (Field 1980, pp. 43–44). We can see how this extrinsic relation can be varied by changing the units of length without affecting the core of the mathematical explanation. And this suggests that we should settle our ontological commitments using only the intrinsic features of the peg/board system. For the intrinsic features are able to explain the phenomena in question, and any additional ontology seems superfluous. This approach commits us to the existence of lengths and their relations, but bypasses the structural relation to numbers. The same strategy came up in Section 3.2.2 in our discussion of Baron's "u-counterfactual theory." For Baron, some mathematical explanations will fail to be genuine mathematical explanations when they have this sort of nonmathematical basis. But as noted there, Baron's specific proposal entails that there are no genuine mathematical explanations if the supervenience principle holds.

So, it seems that the existence of mathematical explanations that have virtues that make them better in terms of generality or robustness than any microstructural physical explanations is not sufficient to establish the explanatory indispensability of any mathematical entities. For one who denies the existence of mathematical entities can simply use the mathematics that we have to provide these mathematical explanations, and yet insist that they are only committed to the existence of the physical properties or relations, like lengths, that are involved.[66] This is a version of what Colyvan calls "easy road nominalism." A nominalist is someone who excludes abstract objects from their ontological commitments. An easy road nominalist proceeds "by admitting the indispensability of mathematics to our best scientific theories, but denying that this gives us any reason to believe in the existence of mathematical entities" (Colyvan 2010, p. 286). Colyvan argues that the easy road nominalist is not actually able to endorse the mathematical explanations found in our best science without also countenancing the mathematical entities that are apparently involved. But this argument has not been very persuasive. One point that Colyvan makes is that the nominalist owes us a nonmathematical version of the mathematical explanations that they endorse: "when some piece of language is delivering an explanation, either that piece of language must be interpreted literally or the

---

[66] See especially Leng (2010, 2020, 2021) for a developed version of this approach.

non-literal reading of the language in question stands proxy for the real explanation" (Colyvan 2010, p. 300). On the literal reading of these mathematical explanations, they involve abstract objects. If one insists on a nonliteral reading, then there must be some nonmathematical proxy at work. Colyvan claims that the only way to restrict one's ontological commitments to these nonmathematical proxies is to provide a nonmathematical theory that specifies their character.

This debate helps to clarify why our earlier discussions of genuine mathematical explanations tended to proceed in such negative terms. For Colyvan and Baker in particular have argued that their genuine mathematical explanations involve abstract objects playing an explanatory role that is significant precisely because in these cases the mathematics is not functioning to simply represent nonmathematical explainers. This is hard to maintain for Putnam's peg/board case, and so other examples have taken center stage, such as the bridges of Königsberg case or evolutionary cases involving cicadas or bees. But, unsurprisingly, the opponents of explanatory indispensability arguments have not been convinced by these cases, as they are all amenable to some kind of nonmathematical reinterpretation. For example, the appeal to graphs in the bridges case can be undercut by supposing that there are "physical graphs" or "p-graphs" specifiable in purely physical terms. Surely, the argument continues, by supervenience, there are such physical structures whenever there is a mapping between a bridge system and an abstract mathematical graph. And so, even in such cases, there is an available proxy that can be specified and used to undercut the explanatory indispensability of any mathematical entities.[67]

The limitations of explanatory indispensability arguments have been sharply described by one of their critics, Saatsi, who complains that "what really matters for the indispensability argument – all that matters! – is whether or not mathematics plays the kind of explanatory role that we should take as ontologically committing" (Saatsi 2016, p. 1051).[68] Our question, then, is how we can use the various notions of genuine mathematical explanation considered in Sections 2–4 to identify this special sort of ontologically committing explanatory role. We need a notion that allows for the supervenience principle and yet still provides for a meaningful distinction between genuine and nongenuine mathematical explanations. In addition, to meet Saatsi's challenge, some of the genuine mathematical explanations need to be sufficient to motivate ontological commitment to some abstract objects. This would be a convincing way to support premise 2.

---

[67] See especially Balaguer (1998) for this sort of argument.
[68] See also Knowles and Saatsi (2021) and Knowles (2021a, 2021b).

None of the monist proposals considered in Sections 2 and 3 seem viable. The pluralist proposal developed by Lange does seem defensible, at least for the special case where the target of the explanation is specified in modal terms. Recall, then, Lange's treatment of the double pendulum case, or the bridges of Königsberg. The mathematical explanation explains why a target must have the features it has, assuming that various contingencies are held fixed. Lange would point out that any appeal to a proxy physical property in this case would not be able to account for the special sort of necessity found in the pendulum or bridge system. So, even if we were comfortable positing a physical proxy for the torus or a family of physical graphs to supplant the abstract graphs, these proxies could not explain the necessity at issue. For, as physical properties, they would only account for relations that are as contingent as physical causes or physical laws. The upshot is that we have at least one way for these genuine mathematical explanations to override the nominalist insistence on avoiding mathematical commitments.

However, the defender of explanatory indispensability arguments should be wary of endorsing Lange's proposal, for it turns out that Lange denies that these explanations commit one to abstract objects like numbers or graphs. In his 2021 article "What Could Mathematics Be for It to Function in Distinctively Mathematical Scientific Explanations?," Lange uses his modal account of such explanations to argue against platonism and in favor of what he and others call an Aristotelian realist interpretation of mathematics (Lange 2021b). The argument takes for granted that each mathematical explanation at issue is an "explanation by constraint" that "explains by providing information about how the explanandum is required by constraints on causal processes and hence is necessary in a stronger way than causal powers could render it" (Lange 2021b, p. 46). It might seem like a platonist interpretation of mathematical objects was consistent with this explanatory role so long as the mathematical facts involving these abstract mathematical objects had the requisite sort of necessity. However, Lange argues that the metaphysical gap between the abstract mathematical objects and the concrete physical systems creates a problem for the platonist. For no matter how necessary the pure math turns out to be, this necessity can only apply to the physical system once the physical system is appropriately related to the abstract objects by some sort of mapping or morphism relation. So, what makes the physical system have the features that it does must be settled prior to this mapping relation obtaining. That is, platonic mathematical objects are not apt to fix how things must be in the physical world.

What we need to make sense of these mathematical explanations, for Lange, is an alternative understanding of what pure mathematics is about. Here Lange defends the Aristotelian realist view that "mathematics concerns mathematical properties possessed by physical systems" (Lange 2021b, p. 50).[69] This bypasses the need for a mapping relation between abstract objects and concrete systems. For the Aristotelian realist, mathematical properties are linked by necessities. This means that if a physical system has some mathematical property, then it must also have some other mathematical properties. So the explanation turns directly on the properties possessed by the physical systems in question. The key difference between the platonist and the Aristotelian realist arises with respect to how various counterfactuals are to be evaluated. For the bridges case, for example, if we ask the platonist what would happen if the abstract mathematical graph were different, then Lange argues that the platonist should admit that nothing different would obtain for the physical bridge system. By contrast, the Aristotelian realist would take the change in the abstract mathematical graph to be a change in some of the properties of the physical bridge system. So, at least in some circumstances, altering a mathematical fact would alter a physical system. This is the key, for Lange, to making sense of mathematical explanations of physical phenomena.

We see then that a defender of premise 2 cannot adopt Lange's account of genuine mathematical explanations without somehow addressing this antiplatonist argument from Lange. More generally, I would argue that the defense of premise 2 requires an account of genuine mathematical explanations that connects the explanatory function of the mathematics to the natures of the mathematical objects involved. Lange's argument shows that even when these natures are resolved by the explanations, there is no guarantee that we will get anything like the platonism the indispensability argument promised.

Interestingly enough, the same point holds for the psychology case that motivated Putnam's discussion of autonomy. Recall that for both the psychological explanations and the mathematical explanations, the acceptance of a supervenience principle led Putnam to locate the subject matter of these explanations in the functional organization afforded by the higher-level structural features of these systems. For psychology, a resolutely functional approach seems to lose what is most distinctive of our mental life, namely the "what it is like" of experience or consciousness. Analogously, a structural or functional approach to mathematical explanations leaves out the intrinsic, nonrelational features of mathematical objects that the platonist places so much emphasis on. The platonist cannot use the role of mathematics in these explanations to conclude more about the nature of mathematics than these explanations would require. If we side with Lange, then we are led to a certain interpretation of the

---

[69] See also Franklin (2008).

nature of mathematics, but the intrinsic character that Lange assigns is inconsistent with platonism.

There may, of course, be some way to think about mathematical explanations so that these explanations *do* involve the intrinsic character of mathematical entities and mandate the existence of some abstract objects. The plausibility of an explanatory indispensability argument will thus turn on the plausibility of thinking of the mathematical explanations in this special way. Baron offers precisely such a proposal in yet another account of genuine mathematical explanations in "Mathematical Explanation: A Pythagorean Proposal" (Baron forthcoming). Baron defines a Pythagorean as a platonist who believes in the existence of abstract objects and who also supposes that some intrinsic mathematical properties of these objects are also possessed by physical objects. On the mathematical side, sometimes a mathematical structure will have such an intrinsic property P that guarantees (or rules out) some mathematical property Q of some aspect of that structure. This allows for a genuine mathematical explanation. Consider a physical system that can take on various configurations or states. Baron's Pythagorean proposal supposes that the physical system can have that intrinsic mathematical property P. Then a state of that system may be guaranteed to possess the intrinsic mathematical property Q. When this relation obtains between global and local mathematical properties, we can explain why the state with property Q is present: this is due to the physical system possessing the global property P. The same point applies when a global mathematical property P prohibits a local mathematical property Q: the physical system's realization of P explains why its states lack Q.

The contrasts between Baron's Pythagorean account and Lange's Aristotelian realism are instructive. Lange argues that the mappings prized by the platonist are inadequate to make sense of how the features of abstract mathematical objects could explain the features of concrete physical objects. So, Lange concludes that the physical objects must possess special mathematical properties. At this point, the core difference between these explanatory mathematical properties and ordinary physical properties is that the mathematical properties stand in necessary relations of a sort that is absent for physical properties. Baron proceeds in a different direction. He supposes that the mappings prized by the platonist *do* involve shared properties between the abstract mathematical objects and the concrete physical objects. So Baron concludes that at least some mathematical properties are shared between the abstract and the concrete. Still, like Lange, for Baron the relations between the properties are the key to the explanatory power of the mathematics. But Baron claims that the properties are intrinsic properties of abstract objects, and so they serve to settle the existence and much of the character of the abstract mathematical objects.

What happens if we combine Baron's Pythagorean proposal with our supervenience principle? Baron claims that his account is consistent with supervenience: "It could be that the mathematical properties supervene on physical ones, but it could also be that some physical properties supervene on mathematical ones. Or there may be no supervenience relation at all" (Baron forthcoming, p. 11). The point seems to be that so long as the mathematical properties are distinct from the physical properties, the existence of genuine mathematical explanations is sufficient for the platonist conclusion. However, Baron is also aware that if supervenience obtains, "an incompatibility between mathematical properties will generally correspond to an incompatibility between subvenient physical properties" (Baron forthcoming, p. 11) and, we might assume, the same correspondence obtains for entailments between properties. This seems to undermine the inference to platonism for the reasons noted in Section 5.1: if there are physical properties that mimic the explanatory relations provided by the mathematical properties, then the determined nominalist will regiment their explanations so they mention only the physical properties. So while Baron's account of genuine mathematical explanation is consistent with supervenience, supervenience undermines the explanatory indispensability argument that Baron seems to want to defend.

Of course, with Lange's Aristotelian realist proposal in mind, another obvious question to ask Baron is why one should posit the existence of abstract mathematical objects in addition to these intrinsic mathematical properties? Baron (forthcoming) does not cite Lange (2021b), but Baron does see clearly how hard it would be for the platonist to use Lange's analysis of mathematical explanations as explanations by constraint in a defense of platonism: "how, exactly, do mathematical objects exert this kind of influence?" (Baron forthcoming, p. 6). Notice that if we drop the abstract mathematical objects from Baron's Pythagorean proposal, we are left with the Aristotelian realist properties that Lange claims are the subject matter of mathematics. Baron's response to this position is direct: "Structural properties on my account make indispensable reference to abstract objects" (Baron forthcoming, p. 25). Baron backs up this assertion by noting some of the difficulties with developing a nominalist account of structural properties without in some way or other quantifying over abstract objects. At this point, though, the debate has shifted away from questions about genuine mathematical explanation to questions about the non-explanatory indispensability of mathematics to our best science. If Baron's Pythagorean proposal is correct, then any nominalist interpretation of mathematics can use that interpretation to provide a reconstruction of the mathematical explanations found in science. For this nominalist interpretation of mathematics will introduce the right sort of structural properties and relations between structural properties.

Looking back, then, at premise 2, we have a better understanding of why it is so unpersuasive to the determined nominalist. We need to address Saatsi's challenge to relate the presence of mathematical explanations to the ontological commitment to abstract mathematical objects. But we have not found an account of these mathematical explanations that avoids a resolute representational reading while also clarifying what it is about these explanations that ties them to the existence of abstract objects. Even if the determined nominalist admits the class of mathematical explanations endorsed by Lange, they still have available an ontology of properties that stops short of commitment to abstract objects. It seems, then, that a platonist interpretation of mathematics in terms of abstract objects is disconnected from the explanatory role of mathematics outside of mathematics.

5.3 Inference to the Best Mathematical Explanation

Colyvan and Baker have often emphasized that an explanatory indispensability argument is directed at scientific realists. Scientific realists typically defend their realism about unobservable entities like atoms and electrons through the use of inference to the best explanation (IBE). So, it is claimed, the scientific realist should also accept the existence of platonic mathematical objects using the very same sort of IBE argument. As Baker puts the point, "A crucial plank of the scientific realist position involves inference to the best explanation (IBE) to justify the postulation in particular cases of unobservable theoretical entities ... the indispensability debate only gets off the ground if both sides take IBE seriously, which suggests that *explanation* is of key importance in this debate" (Baker 2005, p. 225). On this reading, premise 1 reflects an endorsement of an appropriate form of IBE. Then the existence of the genuine mathematical explanations required for premise 2 gets us the existence of abstract mathematical objects. For the very same explanatory considerations that speak in favor of atoms and electrons also speak in favor of numbers and groups.

It turns out to be difficult to articulate and defend a form of IBE that supports a plausible form of scientific realism as well as mathematical platonism. To appreciate the challenges, consider a classic explanatory argument for the existence of God popular from the beginning of the nineteenth century. What explains the existence of the universe and the apparent design that we find in some natural objects like living things? The best available explanation of these facts in 1800 was that God created the universe and is directly responsible for these features of living things through various acts of special creation. So, if we accept IBE, then it looks like we should also accept the existence of God. In schematic form, IBE is a form of inference that can be boiled down to premise (A) and conclusion (B):

A. For phenomena $P_1, \ldots, P_n$, the only available potential explanations are $E_1, \ldots, E_r$, and of these E_1 is the best.

B. E_1.[70]

The goodness of the potential explanations is established by considering how well each explanation exhibits explanatory virtues like simplicity, generality, and depth. The God explanation was superior in these respects to all other available explanations, and so it looked reasonable to conclude that God exists. It is a simple explanation that requires only one entity. It is a general explanation because it covers a number of phenomena. And it is a deep explanation in the sense that it guarantees the phenomena in question by explaining all their salient aspects.[71]

However, the further development of science raises difficult questions about the use of IBE. One reading of this challenge is that the virtues that seemed sufficient to activate IBE were actually not sufficient. That is, scientists who argued for the existence of God in 1800 using IBE were using it incorrectly. Now, with the benefit of history, we can apply IBE more carefully and in line with a better grasp of the virtues that are the right ones to appeal to when determining the best explanation. Consider, first, simplicity. Currently, we explain the origin of the universe by appeal to the Big Bang and we explain the apparent design shown by many living things using Darwin's theory of evolution by natural selection. We can recognize that the God explanation of these two phenomena is simpler than our combination of these two explanations, but we do not take this simplicity to establish that the God explanation is superior in the sense that is relevant to the use of IBE. The same points hold for generality and depth. A more general explanation is one that covers more phenomena, and it is hard to deny that the God explanation, were it true, would cover many more phenomena than our combination of the Big Bang and evolution. Still, this does not mean that the God explanation is better for the purposes of IBE. The virtue of depth pertains to the modal force that the explanation assigns to the target of explanation as well as the level of detail that the explanation achieves. Our accepted explanations using the Big Bang and evolution grant that the targets are highly contingent, and also leave many details unexplained. But, again, this does not call into question our use of IBE in support of these theories, and our rejection of an IBE argument for the existence of God.

One reaction to this historical episode is that our current use of IBE should take some explanatory virtues more seriously than others when they guide inference.[72] Virtues like simplicity, generality, and depth are pseudo-virtues for the use of IBE,

---

[70] See Lipton (2004) for an extended discussion.

[71] This last point assumes a Leibnizian God who always acts for a reason.

[72] A consideration of other reactions is not feasible here.

and we can see this by reflecting on various past scientific failures. Still, this does not settle how we should use IBE now. My suggestion is that a defense of scientific realism requires elevating some explanatory virtues so that they establish a threshold for the use of IBE. Once these thresholds are met, a potential explanation is in the running for support through the use of IBE. But if these special explanatory virtues are not met, then we should refrain from using IBE. A more detailed discussion of this approach to IBE would involve showing that the arguments that we accept, such as IBE arguments for the Big Bang or evolution, conform to these restrictions, while arguments that we now reject, such as IBE arguments for the existence of God, fail to meet these restrictions.

The special explanatory virtues that I have in mind are what I will call conservativeness and modesty. The most conservative potential explanation deploys only entities that we have some prior and independent reason to believe in. X is a less conservative potential explanation than Y when X posits more kinds of new entities than Y does. Consider someone who is agnostic about the existence of God, and yet already believes that the universe has been around for a finite amount of time. For this person, the God explanation of the existence of the universe is less conservative than the Big Bang explanation. This person already accepts the existence of some event that is the start of the universe, but does not already accept anything like God. So this person would be entitled to rank the Big Bang explanation as better than the God explanation in respect of conservativeness. Other things being equal, they could then apply IBE and come to a justified belief in the Big Bang theory.

The second special explanatory virtue to consider is what I am calling modesty. The most modest potential explanation is one that deploys new entities only when all the essential features of those new entities are fixed by the explanation of the phenomena in question. X is a more modest potential explanation than Y when the new entities that X posits have more of their essential features fixed by the explanation than the new entities that Y posits. If we merely consider the existence of the universe, the God explanation and the Big Bang explanation are not very modest because they ascribe a number of features to the entities they posit, and few of these features are fixed by the proposed explanation. For example, as Hume pointed out, for God to create the universe, he need not be all powerful, but only powerful to a certain extent. So if the explanation assumes that God is all powerful, we have a violation of modesty. Similarly, the Big Bang explanation ascribes a very high temperature to the universe in the moments after it began. But if we consider simply the existence of the universe, then this very high temperature is unrelated to the explanation of that existence. So, just as with the all powerful God explanation, we have a violation of modesty.

The key difference between the God explanation and the Big Bang explanation is that the addition of other phenomena besides the existence of the universe is sufficient to fix many of the essential features of the Big Bang, while not fixing the essential features that are traditionally ascribed to God. Perhaps the most significant phenomenon that is used to support the existence of the Big Bang is the cosmic microwave background radiation. This is the fact that there is radiation everywhere in the universe that is responsible for a background temperature of around 2.7 degrees Kelvin. This temperature is explained by the assumption of a specific range of temperatures throughout the universe, from near to its very beginnings up to the present day. So this explanation serves to fix how much energy was present in the moments right after the Big Bang. By contrast, at least if we follow Hume, nothing in the natural world settles whether God must be all powerful or merely very powerful, with a wide range left open. So the proposed God explanation violates modesty to the extent that it assumes an all powerful God, over and above what is needed to explain the natural phenomena.

I suggest, then, that the best defense of scientific realism will invoke a highly restricted form of IBE that requires that a potential explanation meet a threshold of conservativeness and modesty for it to be a candidate for the use of IBE. If this is right, then explanatory arguments for platonism that appeal only to natural phenomena involving the concrete world are in trouble. For consider an agent A who has identified a range of phenomena, but who does not yet believe in the existence of any abstract objects such as the natural numbers. A platonist B may propose some potential explanation involving the natural numbers. Let us suppose that the potential explanation counts as a genuine mathematical explanation according to some viable theory of these explanations. This genuine mathematical explanation will be different from its non-platonist competitors, which we can assume only posit new physical objects, properties, or relations. These differences will go along with a difference in explanatory virtues. Genuine mathematical explanations tend to be simpler, more general, and deeper than their physicalist competitors. So, the platonist B concludes that, so long as person A accepts IBE, they should also accept the genuine mathematical explanation as legitimate and so endorse the corresponding platonist ontology of abstract objects.

The problem with this appeal to IBE is that it ignores the problems from the history of science that motivated the special status of conservativeness and modesty. For the platonist B, who already believes in abstract mathematical objects, there is no issue tied to conservativeness or modesty. There are no new entities in the explanations they are proposing, and so they can endorse these new explanations, just as the theist who already believes in God can use God to explain the existence of the universe. But for person A, who does not believe in abstract mathematical objects, the situation is quite different. There is no way to rationally persuade them to endorse the genuine mathematical explanation because this explanation will fall below the threshold required by conservativeness and modesty. First, in proposing a new sort of object, the mathematical explanation will be less conservative than its physical competitor. Second, it is hard to see how the genuine mathematical explanation will satisfy modesty. Compared at least to the physical explanation, the character of the abstract objects will not be fixed by what is needed to explain the target physical phenomena. This is easy to see for a proposed explanation that involves the natural numbers. It is very unlikely to involve more than a few of the natural numbers. So the genuine mathematical explanation will not draw on or require infinitely many numbers. For this reason, a competing physical explanation will be able to get by using only surrogates that simulate whatever finite structure is needed to explain the target phenomena. Unlike in the Big Bang case, then, we get violations of conservativeness and modesty when we try to use IBE to justify a new belief in the existence of abstract mathematical objects. The challenge to the platonist is to try to find a defensible version of IBE that is central to a defense of scientific realism, while also powerful enough to license the use of IBE needed to add platonic entities to person A's ontology. This challenge has arguably not been met, and this is a way to diagnose why explanatory indispensability arguments remain unpersuasive.

5.4 Conclusion

Much of the current interest in mathematical explanation is tied to the hope that a satisfactory account of genuine mathematical explanation would help to settle the right interpretation of pure mathematics. Our examination of these arguments has been divided into three parts. First, I considered the argument from Putnam that the special sciences like psychology afford a special kind of explanation that is not available in physics. I endorsed this argument for what I called the weak autonomy of mathematics: some mathematical explanations exhibit explanatory virtues that are absent from their nonmathematical alternatives. Then I turned to explanatory indispensability arguments that try to use the special character of genuine mathematical explanations to persuade nonplatonists to become platonists. I argued that these arguments do not work. Weak autonomy is not enough to motivate such a platonist interpretation. The attempt to supplement weak autonomy with a more detailed account of how mathematical explanations work also turned out to be insufficient for the platonist conclusion, as becomes clear in Lange (2021b) and Baron (forthcoming).

Finally, we briefly considered an appeal to IBE in support of platonism. Here I suggested that a defensible form of IBE should privilege virtues like conservativeness and modesty that restrict when IBE can be used. If the scientific realist only endorses this restricted form of IBE, then they have no reason to accept the platonist interpretation of mathematical explanations of physical phenomena.

References

Baker, A. (2005). Are There Genuine Mathematical Explanations of Physical Phenomena? *Mind*, **114**(454), 223–238.

Baker, A. (2009a). Mathematical Accidents and the End of Explanation. In O. Bueno and Ø. Linnebo, eds., *New Waves in the Philosophy of Mathematics*, Palgrave Macmillan, pp. 137–159.

Baker, A. (2009b). Mathematical Explanation in Science. *The British Journal for the Philosophy of Science*, **60**(3), 611–633.

Baker, A. (2012). Science-Driven Mathematical Explanation. *Mind*, **121**(482), 243–267.

Baker, A. (2021). Circularity, Indispensability, and Mathematical Explanation in Science. *Studies in History and Philosophy of Science*, **88**, 156–163.

Balaguer, M. (1998). *Platonism and Anti-Platonism in Mathematics*, Oxford University Press.

Baron, S. (2016). Explaining Mathematical Explanation. *The Philosophical Quarterly*, **66**(264), 458–480.

Baron, S. (2019). Mathematical Explanation by Law. *The British Journal for the Philosophy of Science*, **70**(3), 683–717.

Baron, S. (2020). Counterfactual Scheming. *Mind*, **129**(514), 535–562.

Baron, S. (forthcoming). Mathematical Explanation: A Pythagorean Proposal. *The British Journal for the Philosophy of Science*, 1–29.

Baron, S., Colyvan, M., and Ripley, D. (2017). How Mathematics Can Make a Difference. *Philosophers' Imprint*, **17**(3), 1–19.

Baron, S., Colyvan, M., and Ripley, D. (2020). A Counterfactual Approach to Explanation in Mathematics. *Philosophia Mathematica*, **28**(1), 1–34.

Cain, A. J. (2010). Deus ex Machina and the Aesthetics of Proof. *The Mathematical Intelligencer*, **32**, 7–11.

Coexter, H. S. M., and Greitzer, S. L. (1967). *Geometry Revisited*, Random House.

Colyvan, M. (2001). *The Indispensability of Mathematics*, Oxford University Press.

Colyvan, M. (2002). Mathematics and Aesthetic Considerations in Science. *Mind*, **111**(441), 69–74.

Colyvan, M. (2010). There Is No Easy Road to Nominalism. *Mind*, **119**(474), 285–306.

Colyvan, M. (2012). Road Work Ahead: Heavy Machinery on the Easy Road. *Mind*, **121**(484), 1031–1046.

Colyvan, M., Cusbert, J., and McQueen, K. (2018). Two Flavours of Mathematical Explanation. In A. Reutlinger and J. Saatsi, eds., *Explanation beyond Causation: Philosophical Perspectives on Non-causal Explanations*, Oxford University Press, pp. 231–249.

Conway, J. (2005). The Power of Mathematics. In A. Blackwell and D. MacKay, eds., *Power*, Cambridge University Press. 71–86

Conway, J. (2014). On Morley's Trisector Theorem. *The Mathematical Intelligencer*, **36**, 3. DOI: https://doi.org/10.1007/s00283-014-9463-3.

Craver, C. F. (2014). The Ontic Account of Scientific Explanation. In M. Kaiser and O. Scholz, eds., *Explanation in the Special Sciences*, Springer, pp. 27–52.

Craver, C. F., and Povich, M. (2017). The Directionality of Distinctively Mathematical Explanations. *Studies in History and Philosophy of Science Part A*, **63**, 31–38.

D'Alessandro, W. (2020). Mathematical Explanation beyond Explanatory Proof. *The British Journal for the Philosophy of Science*, **71**, 581–603.

D'Alessandro, W. (2021). Proving Quadratic Reciprocity: Explanation, Disagreement, Transparency and Depth. *Synthese*, **198**(9), 8621–8664.

Detlefsen, M., and Arana, A. (2011). Purity of Methods. *Philosophers' Imprint*, **11**(2), 1–20.

Field, H. (1980). *Science without Numbers: A Defence of Nominalism*, Princeton University Press.

Franklin, J. (2008). Aristotelian Realism. In A. Irvine, ed., *Handbook of the Philosophy of Science. Philosophy of Mathematics*, Elsevier, pp. 101–153.

Gorjian, I., Karamzadeh, O. A. S., and Namdari, M. (2015). Morley's Theorem Is No Longer Mysterious! *The Mathematical Intelligencer*, **37**, 6–7.

Hafner, J., and Mancosu, P. (2005). The Varieties of Mathematical Explanation. In P. Mancosu et al., eds., *Visualization, Explanation and Reasoning Styles in Mathematics*, Springer, pp. 215–250.

Hafner, J., and Mancosu, P. (2008). Beyond Unification. In P. Mancosu, ed., *The Philosophy of Mathematical Practice*, Oxford University Press, pp. 151–178.

Hempel, C. (1965). *Aspects of Scientific Explanation and Other Essays in the Philosophy of Science*, Free Press.

Ingram, V. M. (1957). Gene Mutations in Human Hæmoglobin: The Chemical Difference between Normal and Sickle Cell Hæmoglobin. *Nature*, **180**(4581), 326–328.

Jackson, F., and Pettit, P. (1990). Program Explanation: A General Perspective. *Analysis*, **50**, 107–117.

Karamzadeh, O. A. S. (2014). Is John Conway's Proof of Morley's Theorem the Simplest and Free of a Deus ex Machina? *The Mathematical Intelligencer*, **36**, 4–7.

Karamzadeh, O. A. S. (2018). Is the Mystery of Morley's Trisector Theorem Resolved? *Forum Geometricorum*, **18**, 297–306.

Kasirzadeh, A. (forthcoming). Counter Countermathematical Explanations. *Erkenntnis*, 1–24.

Kitcher, P. (1989). Explanatory Unification and the Causal Structure of the World. In P. Kitcher and W. Salmon, eds., *Scientific Explanation*, University of Minnesota Press, pp. 410–505.

Knowles, R. (2021a). Platonic Relations and Mathematical Explanations. *The Philosophical Quarterly*, **71**(3), 623–644.

Knowles, R. (2021b). Unification and Mathematical Explanation. *Philosophical Studies*, **178**(12), 3923–3943.

Knowles, R., and Saatsi, J. (2021). Mathematics and Explanatory Generality: Nothing but Cognitive Salience. *Erkenntnis*, **86**(5), 1119–1137.

Koslicki, K. (2012). Varieties of Ontological Dependence. In F. Correia and B. Schnieder, eds., *Metaphysical Grounding: Understanding the Structure of Reality*, Cambridge University Press, pp. 186–213.

Kostić, D., and Khalifa, K. (2021). The Directionality of Topological Explanations. *Synthese*, **199**(5), 14143–14165.

Kuorikoski, J. (2021). There Are No Mathematical Explanations. *Philosophy of Science*, **88**(2), 189–212.

Lange, M. (2009). Why Proofs by Mathematical Induction Are Generally Not Explanatory. *Analysis*, **69**(2), 203–211.

Lange, M. (2013). What Makes a Scientific Explanation Distinctively Mathematical? *The British Journal for the Philosophy of Science*, **64**(3), 485–511.

Lange, M. (2017). *Because without Cause: Non-causal Explanation in Science and Mathematics*, Oxford University Press.

Lange, M. (2018a). A Reply to Craver and Povich on the Directionality of Distinctively Mathematical Explanations. *Studies in History and Philosophy of Science Part A*, **67**, 85–88.

Lange, M. (2018b). Mathematical Explanations That Are Not Proofs. *Erkenntnis*, **83**(6), 1285–1302.

Lange, M. (2019). Ground and Explanation in Mathematics. *Philosophers' Imprint*, **19**(33), 1–18.

Lange, M. (2021a). Asymmetry as a Challenge to Counterfactual Accounts of Non-causal Explanation. *Synthese*, **198**(4), 3893–3918.

Lange, M. (2021b). What Could Mathematics Be for It to Function in Distinctively Mathematical Scientific Explanations? *Studies in History and Philosophy of Science Part A*, **87**, 44–53.

Lange, M. (2022a). Challenges Facing Counterfactual Accounts of Explanation in Mathematics. *Philosophia Mathematica*, **30**(1), 32–58.

Lange, M. (2022b). Inference to the Best Explanation as Supporting the Expansion of Mathematicians' Ontological Commitments. *Synthese*, **200**(146), 1–26.

Leng, M. (2010). *Mathematics and Reality*, Oxford University Press.

Leng, M. (2020). Mathematical Explanation Doesn't Require Mathematical Truth. In S. Dasgupta, R. Dotan, and B. Weslake, eds., *Current Controversies in Philosophy of Science*, Routledge, pp. 51–59.

Leng, M. (2021). Models, Structures, and the Explanatory Role of Mathematics in Empirical Science. *Synthese*, **199**(3), 10415–10440.

Lewis, D. (2004). Causation as Influence. In J. Collins, N. Hall, and L. A. Paul, eds., *Causation and Counterfactuals*, Massachusetts Institute of Technology Press, pp. 75–106.

Lipton, P. (2004). *Inference to the Best Explanation*, 2nd ed., Routledge.

Lyon, A. (2012). Mathematical Explanations of Empirical Facts, and Mathematical Realism. *Australasian Journal of Philosophy*, **90**, 559–578.

Machamer, P., Darden, L., and Craver, C. F. (2000). Thinking about Mechanisms. *Philosophy of Science*, **67**(1), 1–25.

Mancosu, P. (2000). On Mathematical Explanation. In E. Grosholz and H. Breger, eds., *The Growth of Mathematical Knowledge*, Kluwer, pp. 103–119.

Mancosu, P. (2001). Mathematical Explanation: Problems and Prospects. *Topoi*, **20**(1), 97–117.

Mancosu, P. (2008). Mathematical Explanation: Why It Matters. In P. Mancosu, ed., *The Philosophy of Mathematical Practice*, Oxford University Press, pp. 134–150.

Mancosu, P. (2018). Explanation in Mathematics. In E. N. Zalta, ed., *The Stanford Encyclopedia of Philosophy*, Summer 2018, Metaphysics Research Lab, Stanford University. https://plato.stanford.edu/archives/sum2018/entries/mathematics-explanation.

Morrison, M. (2000). *Unifying Scientific Theories: Physical Concepts and Mathematical Structures*, Cambridge University Press.

Nahin, P. J. (2004). *When Least Is Best: How Mathematicians Discovered Many Clever Ways to Make Things as Small (or as Large) as Possible*, Princeton University Press.

Nandor, M., and Helliwell, T. (1996). Fermat's Principle and Multiple Imaging by Gravitational Lenses. *American Journal of Physics*, **64**(1), 45–49.

Pincock, C. (2015a). Abstract Mathematical Explanations in Science. *British Journal for the Philosophy of Science*, **66**(4), 857–882.

Pincock, C. (2015b). The Unsolvability of the Quintic: A Case Study in Abstract Mathematical Explanation. *Philosopher's Imprint*, **15**(3).

Pincock, C. (2018). Accommodating Explanatory Pluralism. In A. Reutlinger and J. Saatsi, eds., *Explanation beyond Causation*, Oxford University Press, pp. 39–56.

Povich, M. (2020). Modality and Constitution in Distinctively Mathematical Explanations. *European Journal for Philosophy of Science*, **10**(3), 1–10.

Povich, M. (2021). The Narrow Ontic Counterfactual Account of Distinctively Mathematical Explanation. *The British Journal for the Philosophy of Science*, **72**(2), 511–543.

Povich, M. (forthcoming). A Scheme Foiled: A Critique of Baron's Account of Extra-mathematical Explanation. *Mind*. DOI: https://doi.org/10.1093/mind/fzac019.

Putnam, H. (1975). Philosophy and Our Mental Life. In *Mind, Language and Reality: Philosophical Papers*, Cambridge University Press, pp. 291–303.

Putnam, H. (2010). *Philosophy of Logic*, Routledge.

Reutlinger, A. (2016). Is There a Monist Theory of Causal and Noncausal Explanations? The Counterfactual Theory of Scientific Explanation. *Philosophy of Science*, **83**(5), 733–745.

Reutlinger, A. (2017). Does the Counterfactual Theory of Explanation Apply to Non-causal Explanations in Metaphysics? *European Journal for Philosophy of Science*, **7**(2), 239–256.

Reutlinger, A. (2018). Extending the Counterfactual Theory of Explanation. In A. Reutlinger and J. Saatsi, eds., *Explanation beyond Causation: Philosophical Perspectives on Non-causal Explanations*, Oxford University Press, pp. 74–95.

Reutlinger, A., Colyvan, M., and Krzyżanowska, K. (2022). The Prospects for a Monist Theory of Non-causal Explanation in Science and Mathematics. *Erkenntnis*, **87**, 1773–1793.

Rice, C., and Rohwer, Y. (2021). How to Reconcile a Unified Account of Explanation with Explanatory Diversity. *Foundations of Science*, **26**(4), 1025–1047.

Ronan, M. (2007). *Symmetry and the Monster: One of the Greatest Quests of Mathematics*, Oxford University Press.

Ryan, P. J. (2021). Szemerédi's Theorem: An Exploration of Impurity, Explanation, and Content. *The Review of Symbolic Logic*, 1–40. DOI: https://doi.org/10.1017/S1755020321000538.

Saatsi, J. (2016). On the "Indispensable Explanatory Role" of Mathematics. *Mind*, **125**(500), 1045–1070.

Salmon, W. (1984). *Scientific Explanation and the Causal Structure of the World*, Princeton University Press.

Salmon, W. (1989). Four Decades of Scientific Explanation. In P. Kitcher and W. Salmon, eds., *Scientific Explanation*, University of Minnesota Press, pp. 3–219.

Senior, A. W., Evans, R., Jumper, J. et al. (2020). Improved Protein Structure Prediction Using Potentials from Deep Learning. *Nature*, **577**(7792), 706–710.

Solomon, R. (2001). A Brief History of the Classification of the Finite Simple Groups. *Bulletin of the American Mathematical Society*, **38**(3), 315–352.

Steiner, M. (1978a). Mathematical Explanation. *Philosophical Studies*, **34**, 135–151.

Steiner, M. (1978b). Mathematics, Explanation, and Scientific Knowledge. *Noûs*, **12**(1), 17.

Stewart, I. (2004). *Galois Theory*, 3rd ed., Chapman and Hall.

Taylor, E. (forthcoming). Explanatory Distance. *The British Journal for the Philosophy of Science*, 1–33.

Woodward, J. (2003). *Making Things Happen: A Theory of Causal Explanation*, Oxford University Press.

Woodward, J. (2018). Some Varieties of Non-Causal Explanation. In A. Reutlinger and J. Saatsi, eds., *Explanation beyond Causation*, Oxford University Press, pp. 117–137.

Wussing, H. (2007). *The Genesis of the Abstract Group Concept*, Springer.

Zelcer, M. (2013). Against Mathematical Explanation. *Journal for General Philosophy of Science*, **44**(1), 173–192.

Cambridge Elements ☰

The Philosophy of Mathematics

Penelope Rush

University of Tasmania

From the time Penny Rush completed her thesis in the philosophy of mathematics (2005), she has worked continuously on themes around the realism/anti-realism divide and the nature of mathematics. Her edited collection *The Metaphysics of Logic* (Cambridge University Press, 2014), and forthcoming essay 'Metaphysical Optimism' (*Philosophy Supplement*), highlight a particular interest in the idea of reality itself and curiosity and respect as important philosophical methodologies.

Stewart Shapiro

The Ohio State University

Stewart Shapiro is the O'Donnell Professor of Philosophy at The Ohio State University, a Distinguished Visiting Professor at the University of Connecticut, and a Professorial Fellow at the University of Oslo. His major works include *Foundations without Foundationalism* (1991), *Philosophy of Mathematics: Structure and Ontology* (1997), *Vagueness in Context* (2006), and *Varieties of Logic* (2014). He has taught courses in logic, philosophy of mathematics, metaphysics, epistemology, philosophy of religion, Jewish philosophy, social and political philosophy, and medical ethics.

About the Series

This Cambridge Elements series provides an extensive overview of the philosophy of mathematics in its many and varied forms. Distinguished authors will provide an up-to-date summary of the results of current research in their fields and give their own take on what they believe are the most significant debates influencing research, drawing original conclusions.

Cambridge Elements ☰

The Philosophy of Mathematics